# Our Night Sky

## Edward M. Murphy, Ph.D.

THE
GREAT
COURSES®

**PUBLISHED BY:**

**THE GREAT COURSES**
Corporate Headquarters
4840 Westfields Boulevard, Suite 500
Chantilly, Virginia 20151-2299
Phone: 1-800-832-2412
Fax: 703-378-3819
www.thegreatcourses.com

# Edward M. Murphy, Ph.D.

Associate Professor, General Faculty
University of Virginia

Professor Edward M. Murphy grew up in Oak Lawn, Illinois, in the southwest suburbs of Chicago. For as long as he can remember, he has been interested in the night sky, astronomy, and spaceflight. He still vividly remembers his very first view of the rings of Saturn through a telescope given to him by his parents. As a high school student, he took a class at the Adler Planetarium that set his goal of becoming a professional astronomer.

Professor Murphy received his bachelor's degree in Astronomy from the University of Illinois at Urbana-Champaign in 1989, his master's degree in Astronomy from the University of Virginia in 1993, and his doctorate in Astronomy from the University of Virginia in 1996. After receiving his Ph.D., he accepted a position at The Johns Hopkins University. There he worked on NASA's Far Ultraviolet Spectroscopic Explorer (FUSE) mission, first as a Postdoctoral Fellow and then as an Associate Research Scientist. In 2000, he returned to Charlottesville to join the faculty at the University of Virginia.

Professor Murphy uses radio telescopes and FUSE to pursue his research on the interstellar medium, the thin gas filling the vast space between the stars in the Milky Way. His Ph.D. thesis was completed at the National Radio Astronomy Observatory (NRAO) in Green Bank, West Virginia, using the 140-foot radio telescope to measure the magnetic fields in the Milky Way. He also used the 140-foot radio telescope and FUSE to study the origin and nature of high-velocity clouds, mysterious clouds of hydrogen moving toward and away from the Milky Way at high speed.

At the University of Virginia, Professor Murphy teaches introductory astronomy for undergraduate students, a seminar class on intelligent life in the universe for first-year students, and a seminar class on teaching astronomy for graduate students. He also offers a noncredit evening class for the local community at the historic Leander McCormick Observatory. The University of Virginia named him a Teaching and Technology Fellow in 2002–2003 and an Ernest "Boots" Mead Honored Faculty Fellow in 2003–2004.

Professor Murphy manages the Education and Public Outreach Program in the Department of Astronomy, which includes the popular weekly Public Nights at McCormick Observatory and biannual Public Nights at Fan Mountain Observatory. Through this program, he gives astronomy presentations to people of all ages and interests, from elementary school children to adults, from the general public to professional educators. He has given hundreds of astronomy presentations and night-sky tours to civic and educational groups and appears regularly on the *Charlottesville Right Now* radio program with Coy Barefoot to discuss current events in astronomy and spaceflight.

With his colleague Dr. Randy Bell of the University of Virginia's Curry School of Education, Professor Murphy offers an annual professional development workshop for teachers of grades 4–9 in which teachers not only learn about astronomy and space science but also about the nature of science and scientific inquiry. He has worked with the Science Museum of Virginia to develop five planetarium shows and exhibits and has been an astronomy instructor for teacher professional development programs at the NRAO Green Bank Observatory.

Professor Murphy has been married to his wife, Susan, since 1991. They have two boys, Max and Michael. ■

# Table of Contents

# Table of Contents

# Our Night Sky

**Scope:**

This course offers you a tour of the night sky and the constellations and other objects we can see in the heavens during each of the four seasons of the year. The goal is to give you a foundation for navigating the sky on your own with a pair of binoculars or a small telescope. Along the way, you will also learn about the sciences of cosmology and astronomy and a bit about the mythology of ancient peoples.

The first two lectures in the course provide some introductory information about the sky and its motions. In the first lecture, we look at the constellation Orion and find out where our modern system of constellations and star-naming conventions originated. We also learn how astronomers classify stars and how to use a planisphere, a perpetual star map that will be a useful tool

throughout the course. In the second lecture, we learn why the sky gets dark at night, and we explore the layout and motions of the sky. We also acquire some tips and terminology for observing the sky with the naked eye. Lecture 3 describes the workings of binoculars and various types of telescopes and provides advice for buying a telescope and optimal viewing.

With Lecture 4, we begin viewing objects in the sky, starting with the Moon and the Sun. We learn the phases of the Moon and the types of terrain you might observe on its surface; we also trace the ecliptic of the Sun and see how the Sun's tilt with respect to the Earth's equator causes our seasons. In Lecture 5, we turn to the observation of planets with a telescope. We define the three types of planets and see interesting features of each. In Lecture 6, we look at comets, meteors, and lunar and solar eclipses, in addition to less well-known phenomena, such as sundogs, the circumzenithal arc, the tangent arc, and auroras. Resources are provided to help you determine when some of these phenomena will be visible.

© Jupiterimages/Photos.com/Thinkstock.

In Lecture 7, we learn to identify Ursa Major and Cassiopeia, which are both useful starting points for finding other constellations. We then use these constellations to travel around the northern sky during the different seasons of the year in Lectures 8 through 11. In these lectures, we see different types of stars, including binary stars, red supergiants, and white dwarfs, and trace the life cycles of stars. We also observe galaxies, nebulas, and star clusters and learn about black holes, quasars, and other astronomical phenomena. In the final lecture, we take a quick tour of the southern sky and the Milky Way. My hope is that this course gives you the tools you need to launch your own lifelong explorations of our night skies. ∎

# The Constellations and Their Stars
## Lecture 1

> Although the constellation boundaries have been standardized, the shapes that represent the characters have not been finalized. You're free to draw any figure that you want to connect the stars. Even in historical atlases, the various authors depicted the constellations in different forms.

This course is a sky tour that will offer you information about the **constellations**, their mythology, and the interesting objects we can see in constellations with a pair of binoculars or a small telescope. You will also learn a little about the relevant physics of cosmology and astronomy that you can see in our night sky.

Throughout history, humans have associated familiar patterns in the sky with mythological figures. Such patterns are called constellations, which means "a gathering of stars." We see, for example, Orion the Hunter, a constellation that is visible in the Northern Hemisphere in winter. Our modern constellations are based on 48 Greek constellations described by the ancient astronomer Claudius Ptolemaeus in *Almagest*, a mathematical treatise written around A.D. 150. The constellations that he recorded had been part of the Greek tradition for hundreds of years. In fact, evidence indicates that some of the classical constellations arose in Mesopotamia around 2000 B.C. in the Sumerian and Babylonian cultures. In 1922, the International Astronomical Union adopted a list of 88 constellations based on the 48 classical Greek constellations and a variety of star atlases and star catalogs.

In the constellation of Orion, you may have noticed that the stars have distinctive colors. Rigel, for example, is blue-white in color, while Betelgeuse is red-orange. The color of a star tells us its temperature. Blue stars are the hottest, while red stars are the coolest. Astronomers divide stars into seven main classes based on their surface temperatures. Each of these seven classes is subdivided into 10 parts labeled 0 through 9.

Astronomers have created a unit of distance, called the **astronomical unit** (**AU**), that is used to describe distances between planets. An AU is the average distance from the Earth to the Sun, which is about 93 million miles. To measure distances outside the solar system, astronomers use light travel time. Light travels at 186,000 miles per second and about 6 trillion miles per year. The typical star you see in the night sky is many tens or hundreds of light years away.

A **planisphere** is a perpetual star map that is most accurate at a certain latitude. It has a rotating disk that has printed on it the stars, the Milky Way galaxy, and the constellations. The days of the year are printed on the outer edge of that rotating disk. The sleeve that holds the disk has a window that shows only the visible part of the sky. To set the planisphere, simply rotate it so that the date and time match up. To get the view of the planisphere to match your view of the sky, rotate it so that the direction you are facing is at the bottom. Now that you know some of the constellations and some of the star names, you can use a star map or a planisphere to look for the bright constellations that are visible at particular times of the year. ■

**The Milky Way galaxy. As a year-round feature of our night sky, our home galaxy is one key element in a planisphere.**

**astronomical unit (AU)**: The average distance from the Earth to the Sun; about 93 million miles or 150 million kilometers.

**constellation**: A gathering of stars.

**planisphere**: A perpetual star map.

Dickinson, *Nightwatch*, chaps. 1–3.

Heifetz and Tirion, *A Walk through the Heavens* or *A Walk through the Southern Skies*.

Ridpath, *Star Tales*.

*Skymaps.com.*

Staal, *The New Patterns in the Sky*.

1. How does the sky change during a night? Set a planisphere (or download star maps) for the night sky tonight. What constellations are visible an hour after sunset? Which ones are visible at midnight? Which ones are visible just before sunrise?

2. How does the sky change during a year? Use a planisphere (or download star maps) to picture the night sky tonight at 10 pm. What constellations are visible at the same time, 10 pm, a month from now? Which ones are visible 6 months from now? Which ones are visible a year from now?

**3.** When is your astrological birth sign visible at night? Search along the ecliptic on a planisphere until you locate your birth sign. Rotate the base plate until your birth sign is high in the sky (along the meridian). Find midnight, and see which day lines up with midnight. This is the date when your astrological sign is high at midnight. Is this date close to your birthday? If not, then determine when the constellation is high up at noon? Is this date close to your birthday? Remember that your birth sign (astrological sun sign) is the constellation where the Sun is supposed to be on the date of your birth. Thus, it will be highest when the Sun is highest (at noon). Nevertheless, Western astrology was developed around 500 B.C. Precession and the modern boundaries of the constellations mean that the Sun is not usually in the constellation of your birth sign on your birthday.

**4.** You may want to draw the meridian on your planisphere. The meridian is the line that runs from due north to due south passing straight overhead. It cuts the sky into an eastern half and a western half. To draw the meridian on your planisphere, you will need a ruler and a marker. Line up the ruler with the arrows that mark midnight and noon. The midnight arrow is labeled, and the noon arrow is exactly opposite (be sure to get the correct one). You will know you are correct if the meridian passes through the grommet that marks the north celestial pole.

**5.** You can mark the zenith on your planisphere as well. Recall that the zenith is the point straight overhead. Objects that pass straight overhead have the same declination as your latitude. That is, at 40° north latitude (e.g., Madrid, Philadelphia, Denver, Beijing) the stars that pass straight overhead have a declination of 40° north. To find the zenith, find the point along the meridian that corresponds to a declination equal to your latitude. Rotate the base plate (the plate with the star map on it) until one of the lines that has declination marked on it is under the meridian. Now find the declination equal to your latitude. Mark that spot on the meridian.

# The Constellations and Their Stars
## Lecture 1—Transcript

Welcome to the course Our Night Sky. My motivation for creating this course comes from years of experience showing people the night sky. During these sky tours many people have told me that they would like to know more about the constellations, their mythology, the stars, and the interesting objects that they can see in the constellations in a pair of binoculars or a small telescope.

I think this quote attributed to Thomas Carlyle expresses a feeling shared by many people: "Why did not somebody teach me the constellations and make me at home in the starry heavens which are always overhead and which I do not half know to this day?" My goal is to introduce you to the beauty and the wonder of the night sky, and to give you a basic knowledge needed to feel more comfortable navigating the sky. By the end, I hope that you will more at home in several respects which we'll be combining throughout the course.

The first is, I would like you to know the constellations and their mythology so that season after season when you see them in the sky they become your good friends. I would like you to be able to notice features that you can see with the naked eye, bright stars and their names. I'd like you to be able to look deeper into the sky with telescopes and binoculars. I also want to teach you a little bit about the relevant science and physics of cosmology and astronomy that you can see in our night sky.

After two lectures of introductory material on the sky and its motions, I'll give you advice in the third on using binoculars and telescopes. In the fourth lecture we'll learn about the motions of the Sun and the Moon in the sky and how best to observe them with a telescope. In the fifth and sixth lectures we'll learn about observing the planets through a telescope and then special events that you should watch for while you're out at night. We'll then divide the sky into six parts by location and by season. I will help you find each of the constellations. I'll tell you some of the stories and mythology associated with them. I'll describe what you can see in each constellation with the naked eye, with binoculars, and with a telescope.

So let's begin with the constellations. What is a constellation? Where did our modern constellations come from? Why do we have these figures in the sky? When we look at the sky, it's human instinct to try and find order and patterns in the random placement of the stars. Throughout history humans in different cultures and at different times would seek out familiar figures and patterns in the sky. They'd associate them with a mythological person or animal or object, and in many cases they actually developed stories about these figures in the sky. We call these figures "constellations," which is a word that means a gathering of stars.

Here for example we see the pattern of stars that we call "Orion the Hunter." We'll use Orion for getting a feeling about what a constellation is as well as learning about the stars that make up a constellation. Orion is visible in the Northern Hemisphere winter, which is summer in the Southern Hemisphere, and it dominates the sky in the early evening in January and February. Many cultures have seen the figure of a person in the stars of Orion because it so resembles the stick-figure drawing of a man. You can start by finding the three stars that make up the belt of Orion; three bright starts very close together. Two stars form the shoulders of Orion. Two stars mark his knees or his feet. His belt isn't going straight across; it's tilted because there's a heavy sword hanging off one side. A small group of stars above Orion's shoulder represents his head. His right arm is raised over his head holding a club and his left hand is out in front of him holding a shield.

I've pictured Orion as he appears to observers in the Northern Hemisphere. For observers in the Southern Hemisphere who are inverted compared to those of us in the Northern Hemisphere, Orion will appear standing on his head. With apologies to those of you in the Southern Hemisphere, until the last lecture when we turn to the Southern Sky we'll picture the images of the constellations in this course as seen from the Northern Hemisphere where they originated, so you may have to turn them over in your head.

In classical Western mythology Orion was seen as a great hero and an excellent hunter. This overlay shows us the figure of Orion the Hunter from the 1661 edition of Johannes Bayer's star atlas, *Uranometria*.

There are many stories of Orion's birth and life and death, and in this course I will concentrate on only a few related to astronomy. Both Diana, Goddess of the Hunt, and Aurora, the Goddess of the Dawn, pursued Orion. Just like the Sun and the Moon, the stars rise and set due to the rotation of the Earth. If we put a camera outside on a tripod and take a long exposure of the sky with the shutter open, we'll see the stars leaving a trail as they rise moving from east to west across the sky.

When Orion sets, his bright stars are some of the last to fade, which means that Aurora stays with Orion as long as she can. When Orion finally fades, Aurora weeps and her tears show up as dew on the flowers and the grass in the morning.

Diana's twin brother, Apollo, was against his sister's pursuit of Orion and was angry that she'd been neglecting her hunting duties. So Apollo tricked Diana into shooting Orion in the head with an arrow, either killing him or blinding him. Orion's poor eyesight is represented by the faint stars that mark his head.

We'll revisit the mythology of Orion later in the course as he appears in the myths of many other constellations. The basis for our modern constellations are 48 classical Greek constellations described by the great ancient astronomer Claudius Ptolemaeus in his 13 volume *Almagest,* which was written around 150 A.D. The *Almagest* was a treatise on mathematical astronomy and it was, by far, the most important text on astronomy published before the Renaissance. Most of what we know about ancient Greek astronomy comes from Ptolemaeus' *Almagest.*

In the *Almagest,* Ptolemy explains a geocentric system of astronomy, with a stationary Earth at the center of the universe, and the Sun, the Moon, the planets, and the stars all orbiting around it. His system was good enough that it allowed astronomers to predict where the celestial bodies would appear in the sky at any time or place. It wasn't a perfect system, though. After all, it had the Earth at the center. It was so successful, however, that it wasn't replaced by a better theory until Nicholas Copernicus published his heliocentric theory, a Sun-centered theory, in *On the Revolutions of the Celestial Spheres* in 1543.

Thus, Ptolemy's work dominated astronomy for 1400 years, and in fact, its name comes from *Al Magisti* which means "the greatest." In the *Almagest*, Ptolemy compiled a catalog of 1,028 stars which he organized by constellation. For each star he indicated its location in the constellation, its coordinates in the sky, and an estimate of its brightness.

Ptolemy didn't create these constellations. They had been part of Greek tradition for hundreds of years before Ptolemy recorded them. For example, in the 8th century B.C., Homer in the *Iliad* mentions Orion, the Pleiades and the Hyades (which are all visible in the winter sky), the Bear (Ursa Major or the Big Dipper) which is visible year round in the Northern Sky, and Bootes which is visible in the spring sky.

We'll learn the locations of all these constellations and star clusters later in the course, but Homer tells us the Bear never bathes in the ocean and it's sometimes called "the Wagon" or "the Wain." Homer understood that Ursa Major circles around the North Pole in the sky never rising and never setting. We'll hear more about this in Lecture 7 on the Northern Sky. In the *Odyssey*, Odysseus keeps the Bear on his left in order to sail east. Homer tells us that the Greeks were already aware of using celestial navigation and using the rough location of the North Pole in the sky to guide a boat. Interestingly Homer doesn't mention how the Greeks used these constellations. For that, we must turn to Hesiod, who lived about 150 years later.

Hesiod, around 650 B.C. in his *Works and Days* uses the stars to mark the passage of the seasons, to tell farmers when they should plant or harvest their crops, as indicators of natural weather phenomena such as dry and wet seasons. Interestingly, neither Homer nor Hesiod made reference to the 12 constellations of the zodiac, which trace the path of the Sun in our sky during a year. If the Greeks of Hesiod's day understood the zodiac it would be highly unusual to write a book on the calendar and seasons and not reference the zodiac. Thus it seems unlikely that they understood the significance of the zodiac by 650 B.C.

The first Greek text that we have on the constellations was written by Eudoxus who lived about 390 to 340 B.C. He supposedly learned the constellations from Egyptian priests and introduced them to Greece. He published two works, *Enoptron,* which means "mirror" and *Phenomena,* which means "appearances," but neither of these survive. *Phenomena* was turned into a poem of the same name by Aratus, which does survive and gives us our first guide to the constellations of the Greeks and the stories associated with them.

In his *Phenomena,* Aratus identified 47 constellations including all 12 zodiacal constellations, and he had named six stars. Sirius is the brightest star in the entire sky—it is in Canis Major, the big hunting dog. Arcturus is the second-brightest star in our sky, in the Northern Sky. Capella is the third-brightest star in the Northern Sky. Procyon is the fourth-brightest but it is down in the Southern Sky. Spica is a bright star in Virgo. Protrygeter, also known as Vindemiatrix, is in Virgo. Vindemiatrix was the faintest of these stars but it was probably included because it was used to determine when to harvest grapes.

Thus, by the 4th century B.C., the Greeks had a well-formed system of constellations, including the zodiacal constellations. Where did they come from, though? Did the Greeks invent them after Homer and Hesiod but before Eudoxus? Did they import these constellations from somewhere else as Eudoxus implies? If so, when and where were the constellations invented?

A clue can be found in the positions of the constellations themselves. The Greek texts on astronomy have no constellations around the South Celestial Pole because that part of the sky was invisible from the location where the constellations were invented. For example, the oldest depiction of the classical constellations in the sky is the famous statue Atlas, the Farnese Atlas, in the National Archeological Museum in Naples, Italy. The statue is a Roman copy that dates to around 150 A.D. of a Greek original, which is now lost to us, but it's presumed to be a few hundred years older.

The statue shows Atlas straining under the weight of holding the celestial sphere on his back, and drawn on the sphere are 41 or 42 of the 48 classical constellations. Look closely at the sphere and you'll see a lack of constellations around the South Celestial Pole that is just above Atlas' head where his shoulder is touching the sphere. Whoever invented the constellations didn't create any constellations in this part of the sky because that part of the sky was invisible to them because of their location on Earth. Since that part of the sky was also invisible to the Greeks, they didn't fill in the empty area.

The size of this unmapped region of the Southern Sky can tell us the latitude of the people who created the constellations. Since there are no classical constellations within 36 degrees of the South Pole, the inventors of the constellations must have lived at a latitude of about 36 degrees north, which is marked here on this map. You can see this is too far south to have been mainland Greece and too far north to have been Egypt.

In addition, the constellations are not centered on the modern North Pole in the sky. Due to the rotation of the Earth, the sky turns from east to west around an axis that passes through two poles in the sky, the North and South Celestial Poles. As we'll see later in Lecture 7, both poles are gradually shifting in position in the sky due to a wobble in the Earth's rotation.

The northern constellations are actually centered on a point that was the Earth's North Celestial Pole around 2,000 B.C. This is too old to be the Greek civilization, so again we can rule them out. The Egyptian civilization is old enough but it is in the wrong place, they are too far south. From stone tablets, such as the Mul.Apin tablets, we know that the Babylonians in modern day Iraq had a sophisticated system of astronomy by 700 B.C. It included 12 zodiacal constellations. The Greek zodiacal constellations are remarkably similar to the older Babylonian zodiac. Both included a Bull, Scales, a Lion, Twins, a Sea-Goat, and an archer that was half man and half horse.

The ancestors to the Babylonians, the Sumerians, were flourishing around 2,000 B.C. and they were at the right latitude. It seems that the evidence indicates that some, if not many, of our classical constellations arose in Mesopotamia around 2,000 B.C. in the Sumerian and Babylonian cultures. By 400 B.C., these constellations had been imported to Greece. The Greek astronomers adopted the patterns in the sky and fit their classical stories to these figures, including the zodiac.

Today a constellation is defined as a region of the sky with well-defined boundaries. In 1922 the International Astronomical Union adopted a modern list of 88 constellations. They were based on the 48 classical Greek constellations described by Ptolemy in 150 A.D. and a variety of star atlases and star catalogs since then, including maps of the Southern sky from the 15th to the 17th centuries. The names of the constellations today are the Latin translations of their Greek names.

Belgian astronomer Eugene Delaporte drew up and published rectilinear boundaries for the 88 constellations in 1930. Anything inside the boundary is part of the constellation, and every part of the sky belongs to one constellation or another.

In this close-up of Orion and the surrounding constellations including Orion's large hunting dog Canis Major (with its bright star Sirius) and his small hunting dog Canis Minor, we see the boundaries of the modern constellations. Note also the Winter Triangle, which by contrast is more of a pattern of stars and it spans more than one constellation.

As we back out, we can see the full set of constellations in the early evening winter sky. Now, though the constellation boundaries have been standardized, the shapes that represent the characters have not been finalized. You're free to draw any figure that you want to connect the stars. Even in historical atlases, the various authors depicted the constellations in different forms. For example, in Johannes Bayer's atlas *Uranometria,* Orion is seen facing away from us. While, in Johann Bodes' *Uranographia* he is seen facing towards us.

An asterism is an easily-recognizable pattern of stars in the sky, just like the Winter Triangle that we saw a few moments ago. They're not officially recognized but they are a very useful way to navigate the sky. Another good example is the bright stars of Ursa Major, the Large Bear. In North America, the bright stars in the Bear are known as the "Big Dipper." Three stars form the handle of the Dipper and four stars form the bowl of the Dipper. It looks like a spoon as seen from the side. In England they're known as the "Plough," and in parts of Europe they're known as the "Wagon."

The two end stars of the Big Dipper point to the North Star, Polaris, around which the whole sky rotates. The Big Dipper is a small part of a full constellation called "Ursa Major," the Large Bear, which is seen in this picture. We'll learn more about Ursa Major and how to better see it as a bear in Lecture 7.

This is a good time to explain star names. Not all stars in our sky have a proper name. And in fact, only about 300 of the 6,000 stars visible with the naked eye have a name. In Ptolemy's catalog of over 1,000 stars, no name was given for each star, simply its location in the constellation. For example, in Orion, the bright star Rigel forms Orion's left foot and Ptolemy gives its location as the bright star in the left foot in common with the water. The coordinates for the star are given (1:19:10, 31:30:1) and he lists its brightness (magnitude 1).

In the 8th of the 10th centuries A.D., Arabic-Islamic astronomers working in Baghdad began to compile catalogs of stars. In part, their astronomy was based on Ptolemy's work, which had been translated into Arabic one to two centuries earlier. In some cases, these Arabic astronomers just took the Greek description of the star and translated it into Arabic. For example, in Orion, "the bright star in the left foot" became "*rijl al-Jauza*" (the foot of al-Jauza). Al-Jauza was a goddess that they saw in this part of the sky.

In other cases though, they used ancient Arabic names for the stars. In the 10th to the 13th centuries A.D., these Arabic translations of Ptolemy's *Almagest* reached Spain where they were translated into Latin. And many of these Latin translations retained the Arabic star names.

By the time of the Renaissance a few centuries after the first translations into Latin, the Arabic star names were long established, but down through the ages many mistakes had been made, mistakes in translation, transcription, and identification. Since so much time had passed and so few records exist from that time, it's difficult to trace exactly what happened.

Therefore, our modern star names are a mix of mistranslated and mistaken Arabic names and Greek and Roman names. This explains the prevalence of star names beginning with the letter "*Al*." "*Al*" is equivalent to the article "the" in the English language. For example, in Orion we have Alnitak and Alnilam, nearby is Aldebaran and Algenib, and in the summer sky we have

Altair. Many other Arabic words appear frequently in our star names. Take "d*hanab*," the Arabic word for "tail." The bright star that forms the tail of Cygnus the Swan is Deneb, the bright star in the tail of Leo the Lion is Denebola. A star in the tail of Cetus, the Sea Monster, is Deneb Kaitos, and a star in the tail of Capricornis, the Water Goat, is Deneb Algedi.

I think the best example of this mistranslation over time is how Orion's right shoulder became known as "Betelgeuse." To Ptolemy and the Greeks, it was simply "the bright reddish star on the right shoulder." When Arabic astronomers translated this into the Arabic language, the star became "*yad al-Jauza*," "the hand of al-Jauza" who was a female figure that they saw in this part of Orion.

In the first medieval translation into Latin the Arabic letter for "y" was confused for the letter "b," so "*yad al-Jauza*" became "*bad al-Jauza*." Renaissance astronomers then mistook "bad" to be the assumed Arabic word "bat," or "armpit." So the star became "*bat al-Jauza*," "the armpit of al-Jauza." Additional errors were made over time, so "*bat al–Jauza*" became "Betelgeuse," which is how we know it today.

In 1603, Johann Bayer published an atlas, *Uranometria*. It was the first star atlas to cover the whole sky and it was the most accurate star atlas of its day because the star positions were plotted to a fraction of a degree. It included beautiful constellation figures drawn over the scientific star chart. The work initiated the modern convention of labeling bright stars in a constellation with lower-case Greek letters followed by the genitive or possessive form of the constellation's name. In the reference book that comes with this course, we've listed the constellation names and their genitive cases.

In most constellations, Bayer named the brightest star "alpha," the second was "beta," the third was "gamma," "delta," and on through the Greek alphabet. Unfortunately he broke this rule about as often as he followed it. In many constellations he had other reasons for naming a star "alpha." For example, in Orion, Betelgeuse is Alpha Orionis and Rigel is Beta Orionis even through Rigel is the brighter of the two stars. In some cases such as Ursa Major, the Big Bear in the sky, Bayer labeled the stars in order along the constellation. He started at the end of the bowl of the dipper with Alpha and then went to the next star, Beta, Gamma, Delta, and along through the Big Dipper.

There are many other star-naming systems in use in astronomy and it's not uncommon for a star to have a dozen or more designations from a variety of star catalogs. Other than proper names, the Bayer designations are the most important that you need to know for the naked eye stars. Today, the names of celestial objects must be approved by the International Astronomical Union. You cannot officially purchase the right to name a star. There's no company that's authorized by the International Astronomical Union to sell star names. A number of companies will sell you the right to name a star, but that name won't be recognized by any astronomer, only by the company selling you the star.

In the constellation of Orion, you may have noticed that the stars have very distinctive colors. In fact, Orion is probably the best constellation for seeing the colors of stars in the Northern Sky. You can compare the colors of Betelgeuse and Rigel by rapidly shifting your gaze from one to the other. Rigel is distinctly blue-white in color while Betelgeuse has a red-orange color. By comparison, the belt stars of Orion are blue.

The color of a star tells us its temperature. Blue stars are the hottest, while red stars are the coolest stars. Astronomers classify stars by their surface temperature. They're divided into seven main classes: O, B, A, F, G, K, and M stars. O stars are the hottest stars and M stars are the coolest stars. You can remember the order from hottest to coolest with the saying, "Oh Be A Fine Girl or Guy Kiss Me." Each of these seven classes is subdivided into 10 parts, labeled 0 through 9. So like take A-type stars for example, an A0 star is slightly hotter than an A1 star, which is slightly hotter than an A2 star. Our Sun is a G2 star. Here's a plot, a chart, that shows you some of the stars, bright stars in our sky and their spectral classification, the type of star that they are. Hot O stars are very, very rare. In Orion we can see a few of these O stars. They're typically violet in color or very deep blue in color. Alnitak is an O9 star and it happens to be the brightest O9 star in our sky and it's one of the stars in the belt of Orion. Other bright stars that we've mentioned, such as Sirius, is a blue A-type star, Procyon is a white F-type star and Arcturus is a yellow-orange K0 star. Betelgeuse, that right star in the shoulder of Orion, is a red M1.5 star.

The stars in a constellation such as Orion are not necessarily, or even typically, related to one another, nor do they all have the same distance. Let's spend a minute getting a feeling for the incredible distances that we'll be looking at in this course. Miles and kilometers are useful when we measure the sizes of planets and moons in our solar system, but when we expand out into the solar system, the distances between planets are millions or billions of miles or kilometers. When the distances get that big, miles and kilometers are too small to be useful. Astronomers have created another unit of distance called the "astronomical unit" that we use in our solar system when we describe the distances between the planets. An astronomical unit is the average distance from the Earth to the Sun, which is about 93 million miles or 150 million kilometers. It's easy to remember the distance from the Earth to the Sun. The Earth is one astronomical unit from the Sun.

In this diagram of the inner solar system we see the orbits of Mercury, Venus, Earth, and Mars to their scale sizes, and a one astronomical unit bar is drawn to give the image some scale. As we zoom out, we see the orbits of the outer planets, planets such as Jupiter, which is 5.2 AU from the Sun or Saturn, which is 9.5 AU from the Sun, or out at the edge of the drawing where we see Pluto at 39.5 AU from the Sun.

Although an AU is large, it's not large enough for measuring distances outside the solar system, and for this, astronomers use light travel time. Light travels 186,000 miles every second or 300,000 kilometers per second. In one and a quarter seconds, light covers the distance from the Moon to the Earth. We would say that the Moon is one and a quarter light seconds away. Light from the Sun takes eight minutes and 20 seconds to get to the Earth. When we see the Sun in the day sky, it looks not as it looks now but as it did eight minutes and 20 seconds ago when the light left. If there's a giant flare on the Sun, we won't know of it for eight minutes and 20 seconds. Jupiter is 43 light minutes away. When we see Jupiter in the sky, we're seeing it as it was 43 minutes ago. Pluto is five hours and 20 minutes away.

The most distant spacecraft that mankind has every launched is the Voyager 1 spacecraft, and it's now 16 light hours away from us, in 2011, and it will be 20 light hours away in 2019. If we send a signal to Voyager 1 using radio waves, which travel at the speed of light, it will take 16 hours for the signal to get out to Voyager 1. It will respond right away and its response will come back at the speed of light and take 16 hours to get to us. The roundtrip time from us to Voyager 1 and back again is over 32 hours. Keep in mind that Voyager 1 is the most-distant spacecraft. It's the deepest that we have ever explored into the universe with a spacecraft.

By comparison, it takes light from the nearest star system, Alpha Centauri, about 4.4 years to reach Earth. Thus, the distance to Alpha Centauri is 4.4 light years. A light year is the distance that light travels in one year; it's about six trillion miles or 9.5 trillion kilometers. When we see Alpha Centauri, we see it as it looked 4.4 years ago when the light left. Alpha Centauri is visible in the Southern Hemisphere. It's not visible to most Northern Hemisphere observers. The typical star that you can see in the night sky is many tens or hundreds of light years away.

Not all the stars in a constellation are related to one another. Some are close to us and some are far away. Take Orion, shown here are the distances to some of the brightest stars. Bellatrix is the closest, only 240 light years away. Then comes Betelgeuse, and Saiph, and then we have Rigel, Alnitak, and Mintaka which all have roughly the same distance of about 800 and 900 light years. These stars are likely related to one another, and they were probably born at the same time. Finally, Alnilam is the most distant of the bright stars at 1,340 light years.

If we were to view Orion, or any constellation, from a star many dozens or hundreds of light years from our current location, the constellation would have a very different shape.

Now, once it gets dark outside, you will need a map or a guide to show you your way around the sky and to locate constellations. You can purchase maps of the sky or download them off the Internet or in the resource book provided with these lectures. I have listed some of the Internet sites where you can find them.

Our view of the sky depends on our location on Earth and it changes during the night and during the year. A star map may only be accurate for a certain date and time and location, but for casually observing the sky, these can be loosely defined. A star map might be good for January, in the early evening, at mid-northern latitudes. Thus, you could use it all month long, between, say, sunset and 10:00 p.m. for most of Europe and North America.

In addition to printed star maps, there are perpetual star maps called "planispheres." A planisphere allows us to determine how the sky will look at any date and time. Like star maps, they depend on your location, so most planispheres will have printed on them the latitudes at which they're most accurate. Even if you don't live at these latitudes, as long as you're no more than say 10 or 20 degrees away, they'll still show you what's high up in your sky.

First, let's note the features on the planisphere. It has a rotating disk that has printed on it the stars, the Milky Way Galaxy, and the constellations. The days of the year are printed on the outer edge of that rotating disk. The sleeve that holds the disk has a window that shows us only part of the sky. This is the part of the sky that's visible. Around the edge of that window we have labels, we have horizons, it's labeled as the "horizon." The cardinal directions are shown, north, east, west, and up here is south. The hours of the day are printed along the inside of this blue circle right here.

You may have noticed on your star map that east and west are reversed, and there's a good reason for this on star maps. A terrestrial map is meant to be held down but a star map is meant to be held up, up in the sky like this. Because you hold a star map up in the sky like this, east and west are reversed.

To set the planisphere, simply rotate the planisphere so that the date and time match up. For example, if I want to set my planisphere to see what the sky looked like on January 15 at 10:00 p.m., I would just rotate the disk until January 15 lines up with 10:00 p.m. If we do that you will see Orion high up in the Southern Sky down on this part of the planisphere. Over here on the east side you'll see Leo rising up, and over here on the west side you'll see the great square of Pegasus and Andromeda just setting.

To get the view of the planisphere to match your view of the sky, you want to rotate the planisphere so that the direction you're facing is at the bottom. If I'm facing south, I want to hold the planisphere so that south is at the bottom. If I'm facing east, I want to hold the planisphere so that east is at the bottom. Now the stars visible in the window will match my view of the sky.

A planisphere, or map, is worth your time, so look one over before the next lecture when we'll talk about the motions in the sky. Now that you know some of the constellations and some of the star names, you can use a star map or a planisphere to go outside and look for the bright constellations that are visible this time of the year, and think about the thousands of years of history that are associated with them.

# Seeing and Navigating the Sky
## Lecture 2

The faintest stars that you can see with the naked eye are magnitude 6. With binoculars, you can see down to magnitude 10. And with an eight-inch telescope, you can see magnitude-14 objects under dark skies. The faintest objects that have been seen by the Hubble Space Telescope are about magnitude 30.

In this lecture, we will explore how and why it gets dark at night, and we will learn about the layout and motions of the sky. Looking at an image of Earth from above the North Pole, the planet appears to be rotating counterclockwise. The Sun is on the right-hand side of the image, and the right half of the Earth—the side that is facing toward the Sun—is experiencing day. The left half of the Earth—the side facing away from the Sun—is experiencing night.

One of the first things we notice when we look at the night sky is the huge variety and brightness of stars. A star can be bright in the night sky either because it is nearby or because it is giving off a lot of energy or both. Astronomers describe the apparent brightness of a star using the **magnitude system**, in which a difference of five magnitudes is exactly a factor of 100 in brightness.

When we look up in the sky, it appears as if the Sun, the Moon, the stars, and so on are placed in a bowl stretching over our heads; this bowl is called the celestial sphere. Given that we are looking at the inside of a bowl, it makes sense to measure the positions of objects using angles in the sky. For these measurements, astronomers use units called arc minutes (1/60 of 1 degree) and arc seconds (1/3,600 of 1 degree).

The point on the celestial sphere straight over your head is called the **zenith**. Where the bowl of the sky meets the ground is the horizon. If we use the visible horizon as a reference point, we can describe the locations of objects in the sky by stating their altitude and azimuth. The altitude of an object is the angular distance from the horizon up to the object. The azimuth of an object is a measure of its position relative to true north.

The **celestial equator** is another important point on the celestial sphere. This is simply the Earth's equator projected out into space; it forms a full circle around the celestial sphere and cuts it into a northern half and a southern half. The **meridian** runs from due north to due south and passes straight overhead, dividing the sky into eastern and western halves.

Astronomers define the location of an object in the sky using a coordinate system similar to longitude and latitude on Earth. The east/west position of an object is called its right ascension (RA, measured in hours, minutes, and seconds of time), and the north/south position is called its declination (DEC, measured in degrees, minutes, and seconds of arc with respect to the celestial equator).

**An ideal observing location will have little or no stray light from bright nearby lights, such as homes and stores and street lights, and it will have clear skies and low humidity.**

The stars we see at night change during the year because the night side of the Earth is facing different directions. From Earth, we see this as a result of the shifting position of the Sun in the sky, although as we know, Earth is orbiting the Sun. Given that there are 360° in a circle and 365 days in a year, the Sun appears to move about 1 degree eastward against the background stars every day. This 1-degree shift corresponds to about 4 minutes of time in the sky. In addition to stars and constellations, we can also see planets and **nebulas** in the night sky. ■

## Important Terms

**celestial equator**: The Earth's equator projected out into space; it forms a full circle around the celestial sphere.

**magnitude system**: A way that astronomers describe the apparent brightness of a star. In this system, a difference of five magnitudes is exactly a factor of 100 in brightness.

**meridian**: A circle on the celestial sphere that passes through the north and south points of the horizon, dividing the sky into eastern and western halves.

**nebula**: A cloud of interstellar dust or gas.

**zenith**: The point on the celestial sphere directly overhead.

## Suggested Reading

Millar, *The Amateur Astronomer's Introduction to the Celestial Sphere*.

Ridpath, *Norton's Star Atlas*, chaps. 1–2, 4.

*Sky & Telescope* magazine or *Astronomy* magazine.

## Questions to Consider

1. Where are the celestial pole, equator, and ecliptic located in your sky? Use your planisphere or download a star map and find the locations of these important markers in your sky tonight.

2. Which planets are visible in your sky tonight and how do they appear? Download a star map that shows the locations of the planets in the sky or use an astronomy magazine for this month to determine their locations. Go outside and find them. How close are they to the ecliptic? What time of night will they be highest? Compare their appearance to some bright stars in the sky. Note especially their colors, brightness, and whether they are twinkling.

3. How fast do the planets move through the sky? Find a bright planet or two in the night sky and observe them over a few weeks. Mars moves fast enough that you will detect its motion relative to the background stars in only a few nights. Saturn moves slowly enough that it might take a few weeks before its motion is evident.

# Seeing and Navigating the Sky
## Lecture 2—Transcript

In the first lecture, we heard the story of Orion the Hunter and how his bright stars dim slowly at dawn. In the evening that process is reversed. As it gets dark, Orion's stars are so bright that they're some of the first to appear. Let's explore how and why it gets dark at night and in this lecture we'll also learn about the layout and motions of the sky, and we'll learn more about observing stars, planets, and other objects with the naked eye.

Let's begin by thinking about the transition into night. If we lived on a planet with no atmosphere, it would get dark almost immediately at sunset. But our atmosphere scatters light so it doesn't get dark right away.

Astronomers have defined three kinds of twilight. Right after sunset we have civil twilight, and during civil twilight you can still read outside without artificial light. Only the brightest stars and planets are visible in the sky, so we can see Venus or Jupiter or the bright star Sirius. Next comes nautical twilight. During nautical twilight, many of the bright stars that are needed for celestial navigation become visible, and you can still see the horizon well enough to determine their altitude. Finally, we reach astronomical twilight when there's not enough light to clearly see that horizon. At the end of astronomical twilight the Sun no longer illuminates the sky and the faintest stars and nebula are visible in a telescope. How long twilight lasts depends on your latitude on Earth. Twilight lasts longer at high latitudes and shorter at lower latitudes. At the equator, astronomical twilight ends one hour and 12 minutes after sunset. At higher latitudes twilight can last two hours or more. The three twilights also operate in reverse in the morning hours before sunrise.

Let's think about why night happens. Here's an image of the Earth as seen from above the North Pole. From this vantage point the Earth appears to be rotating counter-clockwise. The Sun is far off on the right-hand side of the image and the right half of the Earth, the side that's facing towards the Sun, is experiencing day. The left half of the Earth, the side facing away from the Sun, is experiencing night.

Let's imagine an observer at four different times of day. At sunset our observer is going from the day side of Earth to the night side of Earth. It gets dark at night because we're passing into the shadow of the Earth, which is stretching off the left side of the picture. At midnight, our observer is halfway between sunset and sunrise and the Sun is on the opposite side of the Earth. At sunrise, our observer is going from night to day, while at noon they're at midday, with the Sun highest in their sky. When night does arrive, how dark it gets depends on where you live. Under exceptionally dark skies in a remote location, about 6,000 stars are bright enough to be seen with the naked eye, but only about half of these are above the horizon and visible at any one time.

Many people who live in developed areas must contend with light pollution. Light pollution is the illumination of the night sky by stray light from human activities. As you can see in this image of a light-polluted sky, the stray light from the city has lit up the distant horizon and the sky. There are no stars visible just above the city, and even high overhead there are fewer stars visible than what you would see under dark skies.

Light pollution represents wasted light, wasted energy, and wasted money. Take a look at this map of the Earth as seen at night. In the map, you can see all the major cities of Earth, the interstate highway system in the United States, the Trans-Siberian railroad in Russia, and the cities along the Nile in Egypt. All the light in this image that you can see is shining up into space. Light shining up into space doesn't do anyone any good. It represents a waste of precious resources while ruining our view of the sky. Light pollution can be prevented by using full cutoff light fixtures that put the light on the ground instead of the sky. Take a look at this image of some streetlights and notice how all the light from the streetlights is shining down on the ground where it's needed, and not into the sky where it's not needed.

An ideal observing location will have little or no stray light from bright nearby lights, such as homes and stores and street lights, and it will have clear skies and low humidity. Higher altitudes also help. But even if you can't observe the sky from a dark location, there are plenty of wonderful sights to see in the sky. If you do go to a remote site away from light pollution, be sure to let somebody know where you are and when you plan to return.

Also make sure that you have permission to use the site because many public areas such as parks close at sunset.

Many people ask me how they can read a star map or a planisphere at night without ruining their night vision. At night, your eye is much less sensitive to red light than to white light, so you can use a flashlight with a red filter. Your eye uses two kinds of light- detecting cells; cones and rods. The cones, which can see in color, are concentrated in the center of your eye whereas the rods are mostly outside of the central area. Rods can't sense color but they're much more sensitive to faint light.

Now, at night we rely primarily on the rods in our eyes for sensing light, and these rods are much less sensitive to red light than to any other color. To prevent the loss of dark adaptation, you can use a dim a red flashlight instead of a bright white flashlight to preserve your night vision. You can purchase red flashlights ready made for astronomy, but it's easy to convert any flashlight into a red flashlight. Simply take a piece of red cellophane or some of that red tape that they sell in automotive stores for repairing broken tail lights on a car and put those over the lens of any flashlight. You can use more than one layer if you need to, to get the flashlight dim enough so that it doesn't ruin your night vision.

A trick to remember when looking at faint objects is to not look directly at them. Since the more sensitive rods are not in the center of your eye but off to the side, overt your vision to focus the light on these more sensitive rods. This will make faint objects look more noticeable. Averted vision can also be used with a telescope or binoculars to make faint objects more noticeable. When looking through a telescope or binoculars, don't stare right at an object, look a little to the side. When you look a little to the side, the light will be focused on your more sensitive rods and the image, the object, will be easier to see.

One of the first things that we notice when we look at the night sky is the huge variety and the brightness of stars. The brightness of a star is represented on a planisphere or a star map by the size of the star symbol. The brighter the star, the larger the symbol. A star can be bright in our night sky either because it's nearby, or because it's giving off a lot of energy every second, or both. The apparent brightness of a star is a measure of how much light we receive from the star. The luminosity of a star, on the other hand, is a measure of how much light or energy that the star is giving off every second. Luminosity is an intrinsic property of the star. It's related to how big the star is and its surface temperature. Alnilam, the middle star in the belt of Orion, is a very luminous O9 star. But because of its distance it's only the fourth-brightest star in Orion. The apparent brightness of a star depends on the luminosity of a star and the distance to the star.

The brightness of an object drops off as the square of the distance to a star. Imagine measuring the amount of light that I get from a star or a source that passes through a small square. This square could represent my eye or it could represent a telescope or it could even represent a solar cell, and we see it here at Position 1. If we move twice as far away from the source, the light that passed through that square is now spread over an area that's twice as wide and twice as tall. The area is four times larger. If I take my eye or my telescope and I move it from Position 1 to Position 2, the light is spread over an area four times larger but my eye and telescope didn't get any bigger, so it intercepts only one- fourth of the light. That is, the object looks one-fourth as bright. Moving twice as far away from a light bulb reduces the brightness by a factor of four, which is two squared. If I move three times farther away from the light bulb, the brightness of the star is reduced by nine times, or three squared.

Astronomers describe the apparent brightness of a star using the ancient magnitude system. The magnitude system dates to the time of the great Greek astronomer Hipparchus. Around 150 B.C. Hipparchus compiled a catalog of nearly 1,000 stars, and he listed their positions in the sky and their apparent brightness. He developed a system for measuring the brightness called the "magnitude system." The brightest stars were first-magnitude stars. The next brightest that he could see were second-magnitude stars and the faintest stars that he could see were stars of the sixth magnitude. It's important to remember with the magnitude system that fainter objects have larger apparent magnitudes and brighter objects have smaller apparent magnitudes. A magnitude 1.0 star is brighter than a magnitude 3.0 star.

Today, the magnitude system is defined so that a difference of five magnitudes is exactly a factor of 100 in brightness. The difference between a first-magnitude star and the faintest sixth-magnitude star, that's five magnitudes difference, is 100 times. The sixth- magnitude star is 100 times dimmer. This means that one magnitude is a difference in brightness of about two and a half times. Two magnitudes are about six times and three magnitudes are about 16 times.

Now that we've quantified and calibrated the magnitude system, when we measure the brightness of objects in the sky it turns out that many are brighter than the first magnitude. These have been assigned magnitudes that are smaller than 1 and they include negative numbers. The bright star Rigel in Orion for example, has a magnitude of 0.12. Sirius, which is the brightest star in our sky, has an apparent magnitude of −1.4. Venus, which is the third-brightest object in our sky, can be as bright as magnitude −4. The full Moon is magnitude −12.7. The Sun is magnitude −26.5. On the other end of the scale, looking at the fainter stars, the North Star, Polaris, is magnitude 2.0. The faintest stars that you can see with the naked eye are magnitude 6. With binoculars you can see down to magnitude 10. And with an eight-inch telescope you can see magnitude 14 objects under dark skies. The faintest objects that have been seen by the Hubble Space Telescope are about magnitude 30.

Whenever you observe the night sky, it's useful to measure the limiting magnitude. The limiting magnitude is an estimate of the apparent magnitude of the faintest star that you can see high over head. An excellent dark sky site will have a limiting magnitude of 6, but a typical urban area with some light pollution will have a limiting magnitude of 4. You won't be able to see stars fainter than the fourth magnitude. A really bright city area might have a limiting magnitude of only 3 with only about 50 stars visible in the sky. There are many factors that can influence limiting magnitude; light pollution is one. It can easily reduce the limiting magnitude by one to two magnitudes or more. Clouds, humidity, or haze can also limit your ability to see faint stars. A bright Moon in the sky will act like light pollution. A full Moon will reduce the limiting magnitude by one to two magnitudes. You should pick your observing nights carefully. Try to avoid nights with a bright full Moon up in the sky and be aware that haze and humidity will limit what you can see. This is especially important if you want to see the fainter, lower-magnitude objects in the sky.

Let's turn our attention now to thinking about how it is that the sky looks over our heads. When we look up in the sky, it appears as if the Sun, the Moon, the stars, the planets, and the Milky Way are placed on a bowl stretching over our heads. We call this bowl the "celestial sphere." Of course the stars are at varying distances from us, some are nearby and some are farther away, but you can't sense that with the naked eye. It looks to us as if they're all painted on the inside of this bowl. Because observers merely looking at the sky have no idea of the actual distances to the objects, we often focus instead on the positions of objects in the sky. Since we're looking at the inside of a round bowl, it makes sense to measure the positions of objects using angles in the sky.

A circle is divided into 360 degrees. A degree is a fairly large unit of measure, and so astronomers have divided a degree into 60 minutes of arc, which we usually abbreviate arcminutes. That's even fairly large, so astronomers have divided an arcminute into 60 seconds of arc, called an arcsecond. That means if there are 60 seconds in a minute, 60 minutes of arc in a degree, there are 3,600 seconds of arc, or 3,600 arcseconds, in one degree.

You can use your hand to measure angular distances in the sky. Your finger is about one degree across. If I take my hand and hold my fist out at arm's length, my fist is about 10 degrees across, whereas if I take my hand and stretch it out, the distance from the tip of my thumb to the tip of my pinkie is about 20 degrees in the sky. This works well for most people because those of us with large hands tend to have longer arms, and so it appears about 20 degrees in the sky.

Both the Sun and the Moon appear about half a degree across in our sky. We've already seen the constellation Orion, and in Orion the belt is just under three degrees across, while the distance from Betelgeuse to Rigel is about 18.5 degrees. The smallest detail that most people can see with the naked eye is one arcminute, and one arcsecond is about the size of a medium-sized coin such as this U.S. quarter, as seen from 3.1 miles away, which is about five kilometers. Imagine how small this quarter would look from three miles away. That's the size of an arcsecond.

The point on the celestial sphere straight over your head is called the "zenith." Where the bowl of the sky comes down and meets the ground is called the "horizon." In practice, the presence of trees and hills and mountains and buildings limits your view of the horizon. Only at sea do you get a perfect horizon. If we use that visible horizon as a reference point, we can describe the locations of objects in the sky by stating their altitude and their azimuth. When we think of the altitude of an object, such as in an airplane, we're measuring how high something is. You can use that to remember that the altitude is how high up in the sky an object appears.

More precisely, the altitude of an object is the angular distance from the horizon up to the object. An object straight overhead has an altitude of 90 degrees. An object down on the horizon has an altitude of zero degrees. So altitude runs from zero all the way up to 90. In this picture, the Sun has an altitude of 60 degrees. It's about two-thirds of the way up in the sky. Sometimes altitude is called "elevation."

The azimuth of an object is a measure of its position relative to true north. You can think of it as if it was a giant compass on the ground. True north is the direction that you would walk to reach the North Pole on the Earth. It's not the same as magnetic north. If you use a compass to determine north, you have to correct the reading of a compass to go from magnetic north to true north. Due north is defined to have an azimuth of 360 degrees or zero degrees. Due east is 90 degrees. Due south is 180 degrees, while due west is 270 degrees, wrapping back around to due north, which is 360 degrees, or zero degrees.

Altitude and azimuth are very convenient ways to describe where an object is located in your sky at this moment in time. Different observers on Earth though have different horizons. So the altitude and azimuth of an object are different for people on different parts of the Earth. The altitude and azimuth of an object also change as the Earth rotates and your horizon shifts in space. Take the Sun on an equinox, when the Sun rises due east and sets due west. At sunrise the altitude of the Sun is zero degrees and the azimuth is 90 degrees. During the morning, the altitude of the Sun increases until it reaches a maximum, around midday, when its azimuth is 180 degrees due south. After that the altitude decreases until it comes back down to zero degrees at sunset when the azimuth is 270 degrees.

Many telescopes operate on an alt-az mount. An alt-az mount moves left and right, or east and west, and it also moves up and down. They're very easy and intuitive to use. If we think about this whole sky over our heads, and not just a bowl over our heads, we picture the full celestial sphere. There are a few important points on the celestial sphere that we need to describe. The first one is the celestial equator. The celestial equator cuts the celestial sphere into two halves: a northern half and a southern half. The celestial equator is simply the Earth's equator projected out into space and it forms a full circle around the celestial sphere. The North Celestial Pole is the point directly above the Earth's North Pole. The South Celestial Pole is the point directly above the Earth's South Pole. Now, when we put our horizon in, we can see only one half of the bowl at any time. The other half of the celestial sphere is hidden from view and it's blocked by the ground.

Your meridian is another line to consider. Your meridian runs from due north to due south and it passes straight overhead. The meridian divides the sky into an eastern half and a western half. We all know that the Sun rises, moves across the sky, and sets. The meridian is a line that you use every day. "A.M." stands for "*ante meridiem*," which is Latin for before midday when the Sun crosses the meridian in your sky. "P.M." is for "*post meridiem*," after midday, after the Sun has crossed this meridian.

The Earth turns from west to east once every 24 hours, and we don't like to think of ourselves as moving, so we usually think of the celestial sphere as rotating around us when, in fact, it's stationary and we're the ones that are moving. The west to east rotation of the Earth makes it look like the sky is rotating from east to west. Here's a movie of the sky during the course of a night, and in the movie we can see the stars moving from east to west. They're rising on the left side of the movie, moving across the sky, and setting over on the right-hand side of the scene. The stars are turning around the South Celestial Pole, which is near the center of the frame, down at the bottom. As the Moon rises, the sky brightens significantly, and though the

Moon isn't visible in the movie, you can see the shadows on the ground shifting as it changes its azimuth across the sky. Finally, with sunrise, the stars disappear.

Instead of a movie, if we placed our camera on a tripod and took a long exposure photograph, we'd see the stars circling the North Celestial Pole over the course of a few hours. The moderately-bright star Polaris is very close to the North Celestial Pole in the sky but not exactly at the pole. On your planisphere, the North Celestial Pole is represented by the grommet at the center of the planisphere. Just as the real sky spins around the pole, your printed sky on the rotating disk spins around the grommet, the North Celestial Pole, at the center of the planisphere.

Astronomers define the location of an object in the sky using a coordinate system similar to longitude and latitude on Earth. The east/west position of an object is called its "right ascension," and the north/south position is called its "declination." Right ascension is often abbreviated "RA" and declination is abbreviated "DEC." Unlike an object's azimuth and altitude, right ascension and declination don't change as the object moves across the sky. Declination is measured in degrees, minutes, and seconds of arc with respect to the celestial equator. The celestial equator is defined to have a declination of zero degrees. Objects north of the equator have positive declinations, while objects south of the equator have negative declinations. Orion for example, is both. It straddles the equator. Betelgeuse is north of the equator and has a positive declination while Rigel is south of the equator and has a negative declination. The North Pole has a declination of +90 degrees while the South Pole has a declination of −90 degrees.

Right ascension, the east/west position in the sky, is measured in hours, minutes, and seconds of time. The starting point for right ascension is the vernal equinox point, a very important point in our sky. For the ecliptic, the path of the Sun in the sky crosses the celestial equator as the Sun goes from the Southern Hemisphere to the Northern Hemisphere on March 20 of every year. This is an imaginary point in the sky. There's no bright star or anything to mark it. It's just where the Sun crosses the equator, and that's the beginning point for right ascension. Right ascension runs from zero to 24 hours.

As an example, let's look at the bright star Betelgeuse. Betelgeuse has a right ascension of five hours and 55 minutes and a declination of +7 degrees 24 minutes. The right ascension means that Betelgeuse will cross the meridian five hours and 55 minutes after the vernal equinox crosses the meridian. The declination means that Betelgeuse is 7 degrees, 24 minutes north of the celestial equator.

The lines of right ascension and declination are drawn on star maps. They're drawn on planispheres as well. Earlier we set our planisphere to 10:00 p.m. on January 15. If you look at the planisphere, you can see the lines of right ascension radiating outward from the grommet, the North Celestial Pole in the sky. Just to the left of Orion you see a right ascension of six hours passing next to Orion. Along this line of right ascension, the declination lines are marked in 10 degree intervals. Notice that the celestial equator passes right above the belt of Orion. Orion's belt is an excellent way to locate the equator in our sky. We can also do declination. Betelgeuse, as we saw, is about eight degrees north. If you look along those lines of declination you can see that Betelgeuse is above the equator, about eight degrees above the equator. Going farther north, we reach the stars of Auriga, about 40 degrees north, and Polaris is, of course, near 90 degrees north.

As the Earth revolves around the Sun, the Sun appears to move through the sky on a path astronomers call the "ecliptic." It passes through a series of 12 constellations that are called the "zodiac." Today it also passes through a 13th constellation, Ophiuchus, that's not part of the classical zodiac. When astronomers defined the modern constellations in 1922, a piece of Ophiuchus cut across the ecliptic, so it is now our 13th zodiacal constellation.

The stars that we see at night change during the year because the night side of the Earth is facing different directions. In January the constellations of Taurus, Gemini, and Cancer are high in the sky. That's because at midnight you're on the side of the Earth facing away from the Sun, so these are the constellations that are high up in your sky. The constellation Scorpius, Sagittarius, and Capricornus are invisible to us because they're in the same direction as the Sun. They'd be up during our daytime when the sky is bright. As the Earth orbits around the Sun, six months later the situation is reversed. In July the night side of the Earth is facing towards Scorpius, Sagittarius, and Capricornus, so they're high up in the middle of the night. During the daytime Taurus, Gemini, and Cancer are up in the sky but invisible because of the bright daylight.

From Earth we see this as a shifting of the position of the Sun in the sky. In January we see the Sun superimposed on the stars of Sagittarius. One month later we see it superimposed on the stars of Capricornus. As we orbit around the Sun we see the Sun shifting position in the sky, not because the Sun is moving but because we're orbiting around it. In March the Sun is against the stars of Aquarius. Since there are 360 degrees in a circle, and 365 days in a year, the Sun appears to move about one degree eastward against the background stars every day. This one degree shift corresponds to about four minutes of time in the sky. Imagine that I go out and see the constellations on January 15 at 10:00 p.m. As you recall from your planisphere, Orion is up high in the sky in the south at this time. The very next night, January 16, the sky will look exactly the same four minutes earlier at 9:56 p.m. due to this one degree or four minute shift in the sky caused by the motion of the Earth around the Sun.

In one month, or 30 days, the sky has shifted by 30 days times four minutes a day, which is a total of 120 minutes, or two hours. The constellations that were high in the sky at 10:00 p.m. on January 15 are now high in the sky at 8:00 p.m. one month later. By March 15, two months later, the sky has shifted by four hours. Now the constellations that were high up at 10:00 p.m. on January 15 are high up around 6:00 p.m. If we look at the sky at 10:00 p.m. on March 15, we'll see that the constellations that were high in the sky, like Orion, are low in the west, getting close to the horizon. If you have a planisphere, set it for 10:00 p.m. January 15 and step through the seasons about 15 days at a time, each one set for 10:00 p.m. and see how the sky changes.

Not only can we see the stars and the constellations in the sky, but we can also see the planets. Since the orbit of the Moon and the orbits of the planets in our solar system are roughly aligned with the Earth's orbit around the Sun, the Moon and the planets are usually found in the zodiacal constellations as well. Going back to our planisphere for a moment, you'll notice that the Moon and the planets are not pictured on the rotating dial and this is because they're constantly moving through the sky. Their locations gradually change from day to day as they orbit the Sun or orbit the Earth in the case of the Moon. We do know, however, that they will always be near the ecliptic. On your planisphere at 10:00 p.m. on January 15 you should see a line cutting across the planisphere about halfway between Orion and Auriga. It passes through Taurus and Gemini, and that line is the ecliptic. The Sun, the Moon, and the planets will always be near the ecliptic in the sky. If you follow the ecliptic over to the right, near the stars of Pisces, you'll see where it crosses the equator and that point is the vernal equinox.

Finding the bright planets in the sky is relatively easy. Only Mercury, Venus, Mars, Jupiter, and Saturn are easy naked eye objects, and they tend to be as bright, or brighter, than the brightest stars. One way to tell the difference between a planet and a star is that planets do not twinkle while stars do. As the light from a star passes through our atmosphere, pockets of air of different temperatures bend the light back and forth. If the light gets bent out of your eye the star blinks off, as it comes back into your eye, the star blinks on, and then off and on and off and on, so the stars appear to twinkle. Planets are much smaller than stars but they're very much closer and they present a much larger image so that their light is not so easily bent into or out of our eye. Planets don't twinkle but stars do.

The best time to see the planets in the sky are dictated by the position of the planet relative to the Earth and the Sun. The planets move through the background constellations as they orbit the Sun. Jupiter, for example, orbits the Sun in just under 12 years. So it takes Jupiter 12 years to go through all the constellations of the zodiac, about one per year. For planets farther from the Sun than the Earth, the best time to see the planet is when it's opposite the Sun in the sky, called "opposition." That's when the Earth is between the planet and the Sun. During opposition, the planet is closest to Earth and will be at its brightest and its largest.

Mercury and Venus, though, because they're both closer to the Sun than the Earth is, are never far from the Sun in the sky. If they're never far from the Sun that means that they spend most of their time above the horizon when the Sun is in the sky, and so they're invisible for a large part of the time. They're only visible just before sunrise in the eastern sky or just after sunset in the western sky. Neither Mercury nor Venus is ever visible in the middle of the night because in the middle of the night, at midnight, we're facing directly away from the Sun looking outward in the solar system. The best times to see Mercury or Venus are when they're as far from the Sun as they can be, and this is called their "greatest elongation." For Mercury it's 28 degrees from the Sun at its best. It never appears more than 28 degrees from the Sun, so this makes Mercury tricky to see. If it's only 28 degrees from the Sun, it will set right after the Sun does or rise just before the Sun does. Venus' greatest elongation is 47 degrees, making it much easier to see. At times of the year, brilliant Venus will be visible for a few hours in the west just after sunset or in the east just before sunrise.

In addition to the stars and planets, there are also fainter, often more extended objects called "nebula," for which you'll often need binoculars or a telescope. "Nebula" comes from the Latin word for cloud. To early astronomers with their small telescopes, not unlike someone using a small telescope today, all these objects other than stars and planets looked like small, faint, fuzzy clouds in the sky. For example, in the sword of Orion we can see that the middle star is not a point of light but a diffused cloud of light. We call this cloud the "Orion Nebula," or "M42." Through a telescope, it's one of the most spectacular objects in the sky. We'll learn much more about the Orion Nebula in Lecture 9.

Some nebulae have been given proper names, like the Orion Nebula, the Lagoon Nebula in Sagittarius, and the Tarantula Nebula. But most nebulae don't have a proper name and they're known by a variety of catalog numbers. The most popular catalog is the Messier Catalog. Charles Messier was a French astronomer who was interested in finding comets, and during his search for comets he turned up other diffuse faint sources in the sky that weren't comets, because unlike comets which orbit the Sun and move from night to night, these objects were remaining stationary. He published a list of these of objects in 1771. The Orion Nebula is the 42nd object in the catalog so it's known as M42.

A more comprehensive catalog, called the "New General Catalog," or "NGC catalog," was compiled in the 1880s and it includes objects from across the sky. Objects in this catalog are known by their NGC number. The Orion

Nebula, M42, is also NGC 1976. You can consult your planisphere and see many of the Messier objects on the planisphere. A planisphere will show things, the bright ones, such as M13, the Globular Cluster, or M31, the Andromeda Galaxy, or M42, the Orion Nebula. However, for most Messier objects, you'll need a telescope and a more detailed star atlas to help you find them in the sky.

Understanding how and why the sky darkens, and the behavior of light, and the range of apparent magnitudes, and the movement of the celestial sphere during the course of a day and night can vastly enrich your appreciation of everything from planets, to stars, to constellations in our night sky.

# Using Binoculars and Backyard Telescopes
## Lecture 3

[Y]ou should always spend many minutes—10 or more—looking at an object through a telescope so that you can wait for that moment when the atmosphere steadies down and you get a nice, clear view.

Binoculars or **telescopes** gather enough light to make faint objects visible. Binoculars are essentially two small lens telescopes mounted side by side, one for each eye. Since a lens telescope typically produces an inverted and left/right–reversed image, binoculars include a series of prisms that turn the image right-side up and left/right–correct. Binoculars are labeled with their magnification and the diameter of the aperture or objective. For example, 7 × 35 binoculars provide 7 times magnification and have an objective lens with a diameter of 35 mm.

The main job of a telescope is to bring light from a distant object to a focus. The focal length is the distance it takes light to come to a focus in a telescope. Telescopes are usually described by the diameter of the lens or mirror (the objective). For example, a 4-inch telescope has an objective lens that is 4 inches or 100 mm across. Telescopes come in three basic types: refractors, reflectors, and catadioptrics. A refractor uses one or more lenses to bring

light to a focus by bending or refracting the light. Reflecting telescopes use one or more curved mirrors to bring light to a focus in front of the primary mirror. The magnification of a telescope is the focal length of the telescope divided by the focal length of the eyepiece: the longer the focal length of the eyepiece, the lower the magnification. High magnifications are

© iStockphoto/Thinkstock.

**A telescope captures significantly more detail than the naked eye or binoculars.**

best used on nights when the atmosphere is steady and the stars are twinkling very little or not at all. The field of view of a telescope is a measure of how much of the sky can be seen in the eyepiece.

Although they tend to be expensive, refracting telescopes offer excellent image quality with high contrast and require little maintenance. One problem with refracting telescopes is chromatic aberration. Reflecting telescopes are less expensive and do not introduce chromatic aberration. They do, however, require more frequent cleaning and adjustment of the mirrors. A catadioptric telescope uses a combination of lenses and mirrors to bring light to a focus. The light path in such telescopes is folded. The light comes in, bounces off the primary and secondary mirrors, and comes out the bottom, which means that the telescope is compact and easy to transport.

**For a first telescope, I think that two things are paramount: ease of use and portability.**

An alt-az mount for a telescope moves to the left and right and up and down, but the viewer is responsible for moving the telescope to track the stars. An equatorial mount is aligned with the rotation axis of the Earth. As the Earth rotates from west to east, making the stars appear to move from east to west, an equatorial mount cancels the Earth's rotation and keeps the stars in the field of view. Thanks to advances in microelectronics, many telescopes today come with computer control and with databases of thousands of interesting objects in the sky. Accessories that help you get the best use from your telescope include a finder scope or a pointing device, eyepieces, a star diagonal, a Barlow lens, and filters.

Many people want to buy the largest telescope they can afford, but this is usually a mistake. Buy a telescope that is easy to transport, quick to set up, and easy to use. ■

## Important Term

**telescope**: An instrument for viewing distant objects; telescopes come in three basic types: refractor, reflector, and catadioptric.

Dickinson, *Nightwatch*, chap. 5.

Ridpath, *Norton's Star Atlas*, chap. 2.

*Sky & Telescope* magazine or *Astronomy* magazine.

## Questions to Consider

1. Which telescopes are most popular right now? Obtain a copy of an astronomy magazine, or browse the Internet, and look at the advertisements for telescopes and equipment.

2. Binoculars will usually have a separate focus on one of the eyepieces to allow each eye to be in focus. Take advantage of this feature, where possible, to be sure that each eye is in focus.

3. How does the cost of a telescope depend on the diameter of the objective?

4. How does the magnification of a telescope depend on the eyepiece used? Why should you purchase low-power eyepieces first? If you own a telescope, calculate the magnification for each of the eyepieces that you use. Record the magnifications on a note card and keep it with the eyepieces for easy reference every time you go observing. You might also calculate or measure the fields of view and record them on the card as well.

5. If you own a telescope, the instruction manual should tell you how to care for the optics and properly align them (called collimation). A reflector will typically require more adjustment than a catadioptric telescope, which will typically require more adjustment than a refractor. If you can, keep a set of tools with the telescope that will allow you to collimate it in the field.

# Using Binoculars and Backyard Telescopes
## Lecture 3—Transcript

In this lecture, we'll learn how binoculars or a telescope can reveal the wonders of our sky. Binoculars or telescopes gather enough light to make faint objects that you normally can't see bright enough so that you can see them. For other objects, such as the Moon and the planets, they'll reveal details that turn these objects into fascinating worlds. After introducing binoculars we'll learn about the different kinds of telescopes, their advantages and disadvantages, and at the end I'll let you know about the accessories that you should have to increase your capabilities to get the most out of your telescope.

Binoculars are an essential piece of equipment for amateur astronomers. They allow you to see objects better than you can see with them with the unaided eye, and some objects actually look better in binoculars than they do through a telescope. In the coming lectures on the constellations, I will take care to note objects that are great for viewing with binoculars, especially those objects that look better in binoculars. A good example of this is the Pleiades, or M45. They're objects that look much better through binoculars. With a good pair of binoculars you can see the whole cluster, but only a few of the stars are visible at the same time in a large telescope.

Binoculars are essentially two small lens telescopes mounted side-by-side, one for each eye. Since a lens telescope typically produces an inverted and left-right reversed image, binoculars include a series of prisms that turn the image right side up and left-right correct.

Binoculars are labeled with their magnification and the diameter of the aperture or the objective. So 7x35 binoculars mean they provide 7x magnification and have and objective lenses with a diameter of 35 mm. Remember that 25 mm is about equal to one inch, so 35 mm is about 1.4 inches. In this lecture, I'll use objective and aperture interchangeably. The objective usually refers to the lens or mirror in a telescope. The aperture usually refers to the opening at the top of the telescope. In all the equipment in this lecture, the diameter of the lens or mirror is the same as the diameter of the opening through which the light passes.

Since the iris in your eye has a maximum diameter of only 7 mm, binoculars collect more light than your eyes can see and they make faint objects look brighter. A 35 mm pair of binoculars collects 25 times as much light as your eye does. Their magnification allows you to see somewhat more detail than you can see with the unaided eye. Some experts recommend purchasing binoculars before you buy a telescope, and there are some advantages to this. Binoculars can be used for other activities such as bird watching or sports. Binoculars will help you become more familiar with the night sky, but I also think there are some disadvantages. The magnification in most binoculars is insufficient to see details when you're looking at the planets. You can't see the rings of Saturn or the moons of Jupiter in most binoculars. And most binoculars have fixed magnification so there's nothing you can do about this.

Typical binoculars in astronomy have magnifications of 6 or 7 times or greater, and apertures of 35 mm or larger. Above 10 times magnification, you may need to use a tripod to keep the images still. A pair of binoculars with image stabilization can raise this somewhat, but higher power binoculars will require a tripod. Binoculars with an aperture of 70 mm or larger can be held by hand for a short period of time, but their weight makes them very difficult to hold for long periods. A heavy duty camera tripod or video tripod will be needed to use a pair of binoculars like this for a long time. Astronomical binocular mounts make positioning and using binoculars like this very easy. You just place the binoculars wherever you want to look in the sky and the mount will hold them there. Many manufacturers make large format binoculars like these specifically for astronomy. Be aware that some of these giant astronomical binoculars may not be able to focus on nearby objects. Giant astronomical binoculars are probably not well suited for bird watching. When you purchase binoculars, be sure to get a pair that can be mounted on a tripod, regardless of their size. And 10x70 or 11x80 binoculars are typical, and without a doubt, these will require a sturdy tripod for long observing sessions

For many people, the real joy is in observing the details of the Moon or the planets, or seeing very faint objects, and this will require a telescope. There are two main reasons for using a telescope. The first is that a telescope collects significantly more light than your eyes, thus making these faint objects look brighter. The second reason is that a telescope also allows you to see significantly more detail than you can see with the naked eye or a pair of binoculars. The main job of a telescope is to bring light from a distant object to a focus. The focal length is the distance that it takes light to come to a focus in a telescope. At this focal point, where the light is brought to a focus, the telescope creates a small image of the distant object. For most telescopes, this image is upside down and/or left-right reversed.

At the focal point, you could put some film to capture the image, or you could put a digital detector like a digital camera to capture that image. But using cameras to photograph the sky involves a whole additional level of equipment and expertise that we won't go into in this course. I've put some references in the resource book that comes with this course for those that would like more information on astrophotography. For that little image we could also put an eyepiece in the telescope and use that eyepiece to examine the image with your eye. An eyepiece is a set of lenses that magnify the image through the telescope. Magnification is not an intrinsic property of a telescope. The magnification, as we'll find, is set by the eyepiece. The same, by the way, is true of binoculars. It's the eyepiece in a pair of binoculars that determines the magnification of the binoculars.

When talking about telescopes, we usually describe them by the diameter of the lens or mirror, called the "objective." A 4-inch telescope has an objective lens or mirror that is 4 inches, or 100 mm, across. The larger the diameter of the objective, the more light the telescope collects. The advantage in buying a larger telescope is that it collects more light. Imagine that it's raining outside and you have a large bucket and a small bucket. The large bucket collects more rain than the small bucket does. The same is true with telescopes. A large telescope collects more light than a small telescope does. Large telescopes also allow you to see more details within the limits imposed by the steadiness of our atmosphere.

Telescopes come in three basic types: refractors, reflectors, and catadioptrics. A refractor is a lens telescope. It uses one or more lenses to bring light to a focus by bending or refracting the light. Reflecting telescopes use one or more curved mirrors to bring light to a focus in front of the primary mirror. This is a very inconvenient place for the light to come to a focus. To observe the image that a mirror telescope forms, we need to get that light out from in front of the mirror and off to the side. You can install a second mirror, called the "secondary," that directs the focused light out of the telescope tube. For example, in a Newtonian reflector, the secondary mirror directs the light out the side, usually up near the top of the tube. In a Cassegrain reflector, the light is actually sent back through a hole in the primary mirror out the back of the telescope. In really big telescopes you actually can work with the light at the focus in front of the telescope. The 200-inch telescope at Mount Palomar in California was large enough that the astronomer could actually sit in a cage at the focal point in front of the primary mirror.

When using an eyepiece in a telescope, the magnification is a measure of how much larger, or closer, the object looks than it would with the unaided eye. A magnification of 100 times means that the object looks 100 times larger, or 100 times closer, than you would see it without an optical aid. The magnification of a telescope is the focal length of the telescope divided by the focal length of the eyepiece. The focal length of the telescope should be listed on the telescope, in the documentation that came with it, or actually printed right on the telescope itself. The focal length of the eyepiece is always printed on an eyepiece, and all modern telescopes have the focal length listed in millimeters.

With this TeleVue–85 telescope right here, we can get different magnifications by using different eyepieces. It has a focal length of 600 mm, which means it takes 600 millimeters for the light to come to a focus. A 50 mm eyepiece will give us a magnification of 600 divided by 50, which is 12 times magnification. A 25 mm eyepiece will give a magnification of 24 times. A 12.5 mm eyepiece will give us a magnification of 48 times, and a 6 mm eyepiece will give a magnification of 100 times. Keep in mind that the magnification goes opposite the focal length of the eyepiece. The longer the focal length of the eyepiece, which means that the eyepiece is usually larger, the lower the magnification. The shorter the focal length and the smaller the eyepiece, the higher the magnification.

The magnification of a telescope can be changed by using different eyepieces, so you should never buy a telescope based on the advertised magnification. In fact, most quality telescope manufacturers never use magnification as a selling point. They know that by changing eyepieces, you can change the magnification of the telescope. Most often, you'll use magnifications between 50 and 300 times. You'll use low magnification far more often than high magnification. Most observers use about 20 to 30 times magnification per inch of objective diameter when observing the planets, but you can and will use higher magnifications on a very good night. You can use magnifications of up to 60 times per inch of objective diameter, but you hardly ever use those magnifications because it's rare that the night air is steady enough to allow us to use a magnification of over 300 times.

In the last lecture, we saw that the twinkling of stars is caused by the light from the star being bent back and forth by pockets of air in our atmosphere. When the light is bent out of our eye, the star blinks off and when it bends into our eye, the star blinks on. On a night when the stars are twinkling wildly, an image through a telescope will appear blurry and indistinct. A planet, for example, will look distorted. Its edges and features will come in and out of focus and it'll appear to swim around the field of view. High magnifications are best used on nights when the atmosphere is absolutely steady and the stars are twinkling very little or not at all. Even on a night of "bad seeing" though, there will be moments when the view through the telescope steadies down and you get a nice, crisp, clear image of the object. For this reason, you should always spend many minutes, 10 or more, looking at an object through a telescope so that you can wait for that moment when the atmosphere steadies down and you get a nice, clear view.

The field of view of a telescope is a measure of how much of the sky you can see in the eyepiece. The larger the field of view, the more sky you will see. Eyepieces that have a large field of view are usually more expensive because they require more lenses, larger lenses, and lenses that are harder to manufacture. There are two ways to determine the field of view. Before you purchase an eyepiece, you can calculate the field of view it will provide. Most eyepiece manufactures will quote the eyepiece apparent field of view. This is the field of view that the eyepiece would give outside of the telescope. The true field of view that you would see through the telescope is the apparent field of view divided by the magnification of the eyepiece.

For example, this 20 mm eyepiece has an apparent field of view of 50 degrees. When used with this telescope with a focal length of 600 mm, it will provide a magnification of 600 divided by 20, which is 30 times. The field of view will be 50 degrees; the apparent field of the eyepiece divided by 30 times magnification, and that gives you an apparent field of view of 1.7 degrees. That's three times the diameter of the full Moon in our sky.

If you don't know the apparent field of view of an eyepiece, you can determine the apparent field of view with observations of the sky. Simply point your telescope at a bright star near the celestial equator, such as any of the belt stars in Orion, Altair, or Spica. Stars on the equator complete a full circle of 360 degrees in 24 hours. Thus, they are moving one degree every four minutes. Position the star just outside the east edge of the field of view of the telescope. If there is a drive, turn it off, and allow the star to track across the field of view as the Earth turns. Time how long it takes for the star to cross the field of view. Knowing that the star moves one degree in four minutes you can determine the field of view of the eyepiece. Make sure though, that the star crosses the whole field of view and not just a corner. For example, imagine it takes two minutes for the star to cross the field of view. If it's moving one degree in four minutes, the field of view would be one-half of a degree.

People often ask, what should I do if I wear eyeglasses? Near- or far-sightedness can be corrected by refocusing the telescope. In this case, you just remove your glasses and focus the telescope. If you wear eyeglasses to correct for astigmatism, you should keep them on while observing. Some manufacturers today are making eyepiece lenses that can compensate for astigmatism. Eye relief is a measure of how far from the eyepiece you need to place your eye to see the view. If you plan to wear glasses while observing, you should buy eyepieces with large eye relief. In general, long focal length eyepieces, the ones with the lowest magnification, will have better eye relief than high magnifications.

There are advantages and disadvantages to each type of telescope. We'll examine the telescopes themselves before we talk about mounts and accessories. For a first telescope, I so think that two things are paramount: ease of use and portability. Refracting telescopes, you'll remember, are those that use lenses to bring light to a focus. The advantages of refractors are that they have excellent image quality with high contrast. The closed tube keeps dust and moisture away from the optics. The beauty of refracting telescopes like this is they require little maintenance, and the optical system itself should never need to be re-aligned to get it to line up with the eyepiece. Only the outer surface of the lens ever needs cleaning and you should rarely, if ever, do that. It's surprising how much dust you can get on the front lens of a telescope and not need to clean it. A typical starter refracting telescope, or lens telescope, will have an objective lens diameter of about two to four inches.

Refractors do have their disadvantages. The first one, the primary one, is cost. For a given aperture or lens, size, refractors are much more expensive than most other types of telescopes. Refractors above six inches in diameter for example, are prohibitively expensive. Another problem with refracting telescopes is chromatic aberration. A lens bends different colors of light by different amounts. Long wavelength red light is bent less than short wavelength blue light, which means it comes to a focus farther away from the lens. With a single lens, you can bring only one color of light to focus at a time, while the other colors of light are out of focus. If you look at a star or a bright object you'll see a halo, or rainbow, around them because of this out-of-focus light. Achromatic telescopes use multiple lenses made of a variety of glasses to significantly reduce, but not eliminate, chromatic aberration. In many quality achromatic telescopes, the remaining color is not noticeable. Apochromatic telescopes use lenses made of exotic materials, such fluorite or extra-low dispersion glass. They can essentially eliminate chromatic aberration, but they can be also be very expensive.

Reflectors are less expensive to make, so they have a low cost per inch of aperture. For a given amount of money, you can purchase a larger reflector than any other kind of telescope. Because the light reflects off of the mirror and doesn't pass through the glass, they don't suffer from chromatic aberration, but they do have disadvantages. Their open tube allows air currents to circulate inside the tube, which may distort your view. The open tube allows dust and moisture into the tube. Reflectors often require frequent adjustment of the mirrors to keep them properly aligned, and in some cases, you need to adjust the mirrors every time you use the telescope. Telescopes with large mirrors are very difficult to transport, and there is a little bit of light lost due to that secondary mirror that blocks some light getting to the big mirror.

A typical starter reflector will have a mirror diameter of four to eight inches. A very popular telescope that provides the lowest cost per inch of aperture is a Dobsonian. A Dobsonian is a Newtonian reflector with a mirror at the bottom and an eyepiece that comes out the side. Dobsonians are usually made of inexpensive materials, such as plywood, and this reduces their cost and makes them a very economical telescope.

A catadioptric telescope, such as this one, uses a combination of lenses and mirrors to bring light to a focus. Most starter catadioptrics are three to eight inches in diameter. This one happens to be 10 inches in diameter. The light path in a telescope like this is folded. The light comes in, bounces off the primary mirror, off the secondary mirror, and comes out the bottom, which means that the telescope is small and compact and easy to transport. Like a refractor, it has a closed tube design. Their disadvantages are that they're more expensive than similarly-sized reflectors. There is a slight loss of light due to the secondary mirror that's in here. Their primary disadvantage though is cost. Catadioptric telescopes like this can be expensive. The two most popular are Schmidt-Cassegrain telescopes and Maksutov-Cassegrain telescopes. Both telescopes work in the same way, but the lens at the front of the telescope has slightly different shapes. Both telescopes perform equally well, and either one would make a good telescope for beginners.

Let's spend a minute now talking about mounts. An alt-az mount, such as we see on this refractor right here, moves to the left and right and it moves up and it moves down. They're very simple and intuitive to use. Alt-az mounts also happen to be inexpensive. Their disadvantage is that they can't track the stars. You'll remember from the last lecture that as stars move through the sky they are moving in both azimuth and elevation, which means that you are responsible for moving the telescope to track to follow the stars. This is easy once you've had some practice doing it, but it can be difficult when you first get started on this, and it can be especially difficult if you have a high magnification eyepiece in the telescope. Some alt-az telescopes come with a computer controlled mount with motors that can actually track the stars for you.

Another kind of mount is an equatorial mount, which is aligned with the rotation axis of the Earth. As the Earth rotates from west to east, making the stars appear to move from east to west, an equatorial mount cancels the Earth's rotation. In an equatorial mount, an axis points at the North Celestial Pole in the sky. The rotation axis of the mount must be well aligned with the rotation axis of the Earth in order for it to cancel the Earth's rotation and keep the object in the field of view. For casual observing, a rough polar alignment will keep the object in the field of view for many minutes, and that's usually sufficient. For more demanding work, like astrophotography, you will need a well-aligned mount so that this axis points precisely at the North Pole in the sky. Now some alt-az mounts, like a fork mount, can be converted to an equatorial mount by including a wedge that simply takes them and inclines them and points them at the pole in the sky.

Thanks to advances in micro electronics, many telescopes today come with computer control for a moderate cost. Most computer-controlled telescopes have a database of thousands of interesting objects in the sky. After an initial alignment you can use the database to find any object above your horizon, and a database of thousands of objects is more than you can probably see in the telescope. With some telescopes, the computer will tell you where to point the telescope to see the object you have selected. Fully computer-controlled telescopes have motors that will slew and point the telescope for you. Computer-controlled telescopes come on both alt-az and equatorial mounts, and both will track an object and keep it in the field. But if you're considering astrophotography with long exposure times, you'll want an equatorial mount.

Most computer controlled telescopes require a quick alignment. They need to know where they're pointing in the sky initially, and they need to know their latitude, longitude, and the current time. They need to be level and they need to know which direction is north. Other telescopes can do this alignment automatically by using GPS satellites and sensors, and they can use cameras that are actually built into the telescope to finish the alignment process.

Now that we've talked about telescopes, their advantages and disadvantages, let's talk about the accessories that you need to make the best use out of your telescope. An essential component to all telescopes is a finder scope or a pointing device. A finder scope is a low-power telescope that allows you to find objects in the sky. Once it's aligned with the main telescope, you can use it to find objects. The main telescope typically looks at such a small part of the sky that you would have to scan back and forth to find the object. If the finder scope is aligned with the main telescope, if you can center the object in the finder scope, it will be centered in the main telescope. In fact, most finder scopes include a small set of crosshairs to help you locate the center of the field.

Short focal length telescopes usually don't require a finder scope. A wide field, low- power eyepiece will give you a wide enough field to find objects. With experience, you can usually determine where a telescope is pointing by sighting along the tube. But many telescopes come with a sighting device that projects a little red dot or a green laser beam up into the sky to indicate where the telescope is pointing, and these often help you as you navigate around the sky. Both finder telescopes and sighting devices must be aligned with the main telescope at the start of every observing session. At the beginning of the night, point the telescope at a distant light or a very bright star. Center the object in the telescope and use the settings on the finder scope or the sighting device to center the object up. Now the finder and the telescope are aligned with one another, but you may have to do this on every observing session.

The next most-important accessory is eyepieces. Eyepieces come in three standard barrel sizes, 0.965, 1.25 and 2.0 inches. The 1.25 inch and 2 inch eyepieces are the most common. They come in focal lengths that range from 4 mm, which are very high magnification, very small eyepieces, up to larger than this one, 55 mm eyepieces, which are very low magnification and have a very large field of view. Remember, the smaller one is high magnification, the larger one is low magnification. High quality eyepieces provide significantly better views than inexpensive eyepieces.

Short focal length eyepieces have high magnification. The advantage to high magnification is that they may show more detail, but the maximum magnification that you can use is limited by the steadiness of the Earth's atmosphere. If the seeing is bad, high magnifications will not show more detail, and they may, in fact, make it look worse. It's rare that the atmosphere is steady enough to allow us to use more than 300 to 400 times magnification. The maximum magnification is also limited by the diameter of the objective of the telescope. The larger the objective, the smaller the details you can see and the higher the magnification you can use. A rough rule of thumb states that you can expect a maximum useful magnification of about 60 to 80 times per inch of aperture. High magnification is most useful for observing the planets in our solar system and for observing double stars where the two stars are so close together the high magnification is required to separate them.

High power eyepieces like this have their disadvantages though. Their field of view is small, the effects of bad seeing are magnified, and any vibration in the mount due to wind or shaking of the ground is much more noticeable. You will use low power eyepieces far more often than high power eyepieces. When it comes time to buy eyepieces, buy the low power eyepieces first.

A star diagonal makes it much more convenient to view objects through the telescope. It contains a mirror or a prism and it simply bends the light at a right angle. A star diagonal is an essential piece of equipment, especially when you're looking at something high in the sky where the eyepiece might be pointing right at the ground. By putting the star diagonal in, you can bring the light out the side of the telescope.

A Barlow lens is a simple concave lens that increases the focal length of the telescope thereby increasing the magnification of an eyepiece, and it too is an essential accessory. A 2 times Barlow will double the effective focal length of the telescope and thereby double the magnification of any eyepiece that you put in the Barlow. A single 2 times Barlow will double the magnification of every one of your eyepieces. Thus it doubles the number of eyepieces that you have. It's important to buy a quality Barlow. An inexpensive or low-quality Barlow will degrade the view. Barlows have the additional advantage that they preserve eye relief of low-power eyepieces, so if you use eyeglasses you can use a low-power eyepiece to get that good eye relief and put the Barlow in to get higher magnification. But don't forget to take it into account when you're buying eyepieces. A 24 mm eyepiece, when used with a 2 times Barlow, will act like a 12 mm eyepiece, so there's no need to go out and buy a separate 12 mm eyepiece.

There are many filters that you can buy for your telescope, and most filters thread into the bottom of the eyepieces using standard threads. The first one is a Moon filter. Especially when the Moon is full, it can be very, very bright and almost blinding through a telescope. A neutral density Moon filter can cut the brightness down to a manageable level. It doesn't change the color of the Moon. It just simply reduces the amount of light. Some Moon filters have variable polarizing filters that allow you to actually adjust the amount of light that makes it through the filter. By rotating the two components you can adjust it from being almost clear to almost opaque.

Another useful filter is a light pollution reduction filter. A light pollution reduction filter can be used in urban areas the cut down the amount of light pollution as seen through the telescope. Some astronomical objects, but not all of them, and especially nebula, are good for using light pollution reduction filters on. They emit only a few colors of light. Most light pollution, on the other hand, emits all colors of light. The goal of a light pollution filter is to let through only those colors that are emitted by the nebula and not all the other colors that are created by light pollution. Stars give off all colors like light pollution, and so it will reduce the stars just as much as it reduces the light pollution. Color filters are another useful accessory when observing the planets. In the next lecture, when we discuss observing the planets through a telescope, I'll talk about how certain color filters can bring out certain features on the planets.

Let me close by saying that many people want to buy the largest possible telescope they can afford, and we call this "aperture fever." Experience tells me that this is a mistake. A large 10-inch Schmidt-Cassegrain telescope requires two people to carry and two people to set up. It can take 30 minutes or more to set it up, 15 to 30 minutes to get it aligned on the sky, and packing it up will take another 30 minutes. On a beautiful night, it's not the kind of telescope that you can use to spend just a few minutes looking at the Moon or the planets. Buy a telescope that is easy to transport, quick to set up, and easy to use. Even after you buy a much bigger telescope in the future to show you those things that can't see through the small telescope, you'll find yourself going back to that small telescope time and time again.

In the next lecture, we'll talk about the motion of the Sun and the Moon in our sky, and how to best see them through a telescope.

# Observing the Moon and the Sun
## Lecture 4

> I think a project that everyone should do at least once in their lives is
> go out and observe the Moon every night for a month. Start with a new
> Moon or a waxing crescent Moon in the early evening. Draw the shape
> of the Moon and note its location in the sky, especially relative to where
> the Sun set. Every day thereafter, go out at about the same time, draw
> the Moon and how it appears in the sky, and do this for a full cycle
> of phases.

You can observe the phases of the Moon by going outside a little after sunset every day and facing west. Over the course of about two weeks, the Moon goes through the **waxing** gibbous phases until you finally see the full Moon, which rises just as the Sun is setting. The phases of the Moon are determined by how much of its lit side we see as it orbits the Earth. When and where we see the Moon in the sky are related to its phases. The new Moon is in the same direction as the Sun, which means that we will not see it at night because it is up when the Sun is up. But it also means that it will rise with the Sun; it will be highest at noon when the Sun is highest; and it will set with the Sun at sunset.

**Tracking the location of the rising Sun is probably humanity's earliest calendar.**

The far side of the Moon is never visible from Earth. The fact that the Moon always keeps the same face toward us means that it must rotate. The Moon rotates on its axis in the same length of time it takes to revolve around the Earth—about a month. The best time to observe the Moon through a telescope is around first or third quarter, and the best place to observe it is along the day/night line (the terminator). The surface of the Moon consists of two types of terrain: cratered highlands, formed by impacts with meteoroids, asteroids, or comets, and the maria, which are dark, flat areas caused by lava flows.

As we learned in Lecture 2, the Sun appears to move through the sky on a path that astronomers call the **ecliptic** as the Earth revolves around the Sun. Every year, it passes through the 12 constellations of the zodiac and a 13th constellation, Ophiuchus. From the Earth, we see this as a shifting of the position of the Sun in the sky. Since the Earth orbits the Sun in a year and it takes 365 days for the Sun to complete one circuit of the sky, which is 360°, the Sun appears to move about 1 degree each day against the background stars. On a celestial sphere, we can see that the Sun's path across the sky is tilted by about 23.5° with respect to the Earth's equator. This tilt is what causes the seasons on Earth.

A common misconception is that the Sun always rises due east and sets due west, but this occurs only on the vernal and autumnal equinoxes, when the Sun is on the equator. The motion of the Sun north and south of the equator is responsible for the changing length of our days. Another common misconception is that the Sun is straight overhead at noon every day. In fact, unless you live between 23.5° south latitude and 23.5° north latitude, the Sun will never be straight overhead.

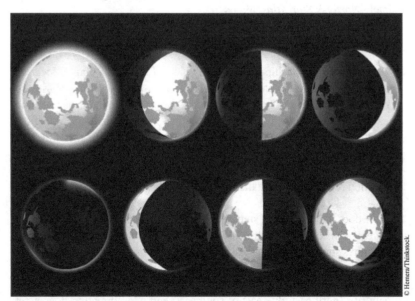

© Hemera/Thinkstock.

**The phases of Earth's Moon. Follow the images from left to right to see the monthly cycle of Moon phases.**

You should never look at the Sun without eye protection, but you can safely view it without a telescope using solar glasses, by looking through number 14 welder's glass, or by projecting the image with a sunspotter or a solar scope. You can also view the Sun through a telescope that has a solar filter or through an H-alpha telescope, which filters out all colors of light coming from the Sun but one. ■

## Important Terms

**ecliptic**: The path on which the Sun appears to move through the sky as the Earth rotates around it.

**waxing**: The phase of the Moon during the transition from new to full.

## Suggested Reading

Dickinson, *Nightwatch*, chap. 8.

Ridpath, *Norton's Star Atlas*, 49–64.

## Questions to Consider

1. What is the relationship between the phase of the Moon and its location in the sky relative to the Sun? Starting with the new moon phase, go outside every clear night for one month (preferably around the same time each night) and draw a quick picture of the phase of the Moon. Note the location of the Moon relative to where the Sun set. After it is full, you may have to get up early in the morning to see the Moon. How does the phase of the Moon change during the month? How is it related to its location relative to where the Sun set or rose?

2. How does the position of sunrise or sunset change at your location during a year? On, or near, each of the four seasonal marker days (the spring equinox, summer solstice, autumnal equinox, and winter solstice) note the location of sunrise and/or sunset relative to landmarks along your horizon (a tree, a neighbor's house, and so forth). What direction does the Sun rise (northeast, due east of southeast) and what direction does it set (northwest, due west, or southwest)? How long is the day?

3. As seen from your location, how does the maximum height of the Sun vary during the year? On each of the seasonal marker days, note how high the Sun is around noon (or 1 pm if you are on daylight saving or summer time). Measure the length of the shadow of a flag pole or other vertical pole. When do you expect the shadow to be longest? When will it be shortest? Will there be a day when it has no shadow?

4. What path does the Sun take through your sky on the summer and winter solstices? From the rising and setting points and maximum heights measured above, you should have a good idea of the path of the Sun on the summer and winter solstices. To see it better, use your planisphere to see the path of the Sun through the sky on the summer and winter solstices. A) On June 21, the Sun is at 6 hours right ascension and $+23.5°$ declination. This is near M35 in the constellation Gemini. Find this location on your planisphere and spin the wheel until that location is on the eastern horizon. This is the moment of sunrise. Note the location of sunrise (northeast, due east, southeast) and note the time of sunrise (find June 21 on the dial and estimate the time that lines up with June 21). Now spin the baseplate and watch the Sun move across the sky. At sunset, note the location and time as well. Calculate the length of the day. B) Now, go through the same exercise for the December 21st when the Sun is at 18 hours right ascension and -23.5° declination. On this day, it is near M8 in Sagittarius. Again, note the sunrise and sunset locations and times, and calculate the length of the day.

# Observing the Moon and the Sun
## Lecture 4—Transcript

The two most familiar objects in our night sky are the Sun and the Moon, and in this lecture we'll talk about both of them. They're the two astronomical objects that enter our lives on a daily basis and we can learn from both of them in a single day. The Moon's monthly phases are one of the most recognizable changes in our sky, and they give us an ever changing view of the fascinating geology of our nearest neighbor. After covering the Moon, we'll move on to talking about the Sun and watching its annual motions through our sky, which is one of mankind's oldest astronomical activities. Its annual motions served as our earliest calendars and were vital to reckoning the seasons and the year. Finally, I'll let you know how to safely observe the Sun through a telescope. It's the only star in the whole universe that we can see up close.

Many of us have had the experience of going outside and seeing a giant full Moon rising in the early evening. Many people, when they see that giant Moon rising up, believe that that rising Moon is actually larger than when the Moon is high over head. But this turns out to be an optical illusion. The Moon is actually slightly larger when it is high overhead than when it's down on the horizon. This illusion is pictured here. The Moon on the very distant horizon appears larger, but if you were actually to take a ruler up to the screen and measure both of those, you'd find out that both Moons are the same size. This illusion isn't very well understood by psychologists but we can certainly measure the size of the Moon as astronomers and we find that it's actually smaller when it's down on the horizon.

What you may not have realized when looking at this giant full Moon rising is that it was probably rising at sunset. In fact, the full Moon always rises at sunset, and why is this? In order to understand that, we have to understand the phases of the Moon. To understand the phases of the Moon, I think the best way to do it is to go outside and observe the phases for two weeks. What we'll do is start at a new Moon—you can always use a calendar to determine when the new Moon will occur—and we're going to go out the same time every day just a little bit after sunset and we'll face west in the sky.

The first thing we'll see on the new Moon day is probably no Moon in our sky at all because that is the definition of a new Moon. The Moon is starting its phases again. So the second day we go out right at sunset again and we face west, and this time we'll see a very thin crescent hanging just above the lit horizon where the Sun had just set. If we go out a day later at sunset again and face the same direction west, we'll see a slightly thicker crescent Moon in the sky, a little bit farther away from where the Sun set. Day after day we will see the crescent moon growing larger and larger and larger. We call these the "waxing phases of the Moon" because it's the way we build a candle. You take a string, dip it in the wax, and pull it out, and as you dip the string in the wax it grows ever larger. The same thing is what the Moon is doing now; the Moon is growing larger over time.

One week after the new Moon you will get to the first quarter Moon when the Moon appears half lit in the sky. After first quarter Moon, the Moon will be more than half lit. It will be in Gibbous phases. These are the waxing Gibbous phases because the Moon continues to grow larger and larger until finally we reach the full Moon. The full Moon will be rising just as the Sun is setting. If we were to go out the following day, right at sunset, the same time we've been going out for these two weeks, we wouldn't see the Moon at all. In order to see the Moon at that time, we need to go out in the morning sky. If we face east just before the Sun rises and now follow the phases of the Moon for two more weeks, we will see that right at sunrise that full Moon is setting in our sky. Over the next week it will go from being a full Moon to getting ever smaller day by day, until finally it's a third quarter Moon a week after it was full. We call these the "waning phases of the Moon" because they're getting smaller and smaller in the sky.

After third quarter we get the waning crescent where it gets ever smaller, until finally we come back to the new Moon again, almost a full month after we started. We've watched the Moon for two weeks in the evening and two weeks in the morning, and from that we can tell that it takes the Moon about a month to go through its phases. More precisely, it's 29.5 days for the Moon to go from new Moon back to new Moon again. In fact the length of our month used to be related to the time it took the Moon to go through its phases. The root of our word "month" derives from Moon. In some calendars today, such as the Jewish and Islamic calendars, the lengths of the months are still related to the phases of the Moon. In both of these religious calendars, the month begins with a new Moon and ends with the next new Moon.

This raises the question, what causes the phases of the Moon in our sky? Why do we see this ever-changing face of our nearest neighbor? The phases of the Moon are the result of how much of the lit side of the Moon we see as it orbits the Earth. Let's take a look at this diagram to see how it works.

On this diagram the Sun is far off on the right-hand side and it's illuminating both the Earth and the Moon. Let's look at what happens during the new Moon. During the new Moon in the sky, the sunlight is coming from the right-hand side. An observer standing on Earth, when they look up at the Moon in the sky, is seeing the unlit side of the Moon. That means that if we're seeing the unlit side of the Moon, there's nothing to see. The lit side is facing away from us. Remember, the Moon is half lit up. It's just that the lit side is facing away from us. A couple of days later, the Moon has moved from Position 1 to Position 2. At Position 2 it's moved enough that now we can see just a little bit of the lit portion of the Moon around the right-hand side. This is the waxing crescent phase. It's "waxing" because it's getting bigger, and "crescent" because it appears as a thin crescent in our sky. As the Moon continues to move it will get gradually larger until we've reached the first quarter Moon when it appears half lit and half dark. You might now ask, why is it a quarter Moon if it's half lit and half dark? The reason is you can see that the Moon is one-quarter of the way through its phases. It's one-quarter of the way through its orbit around the Sun.

After the first quarter phase, we get to the waxing Gibbous phases. After first quarter, the Moon continues to move around, and a few days later when we view the Moon, we're seeing it more than half lit in our sky. If you look very carefully you will see that there is a thin crescent of darkness on the side of the Moon indicating that it's not quite yet a full Moon. These are the waxing Gibbous phases. "Gibbous" actually is an old word that means hunchback because the Moon is a little bit more than half full. Finally, two weeks after this began, we see the full Moon in the sky when the Moon is opposite the Sun. The full Moon occurs because we're looking at the fully-illuminated portion of the Moon. The unlit side of the Moon is facing away from us. After full Moon, the phases reverse their order. We go to the waning

Gibbous phases, to third quarter, the waning crescent phase, and back to new Moon again.

A very common misconception is that the phases of the Moon are caused by the shadow of the Earth. Many people think the crescent Moon is actually caused by the Earth's shadow falling on the Moon. But look at this diagram again. Remember the Sun is off on the right-hand side, so the shadow of the Earth is moving off to the left. When we see the waxing crescent Moon in the sky, the waxing crescent Moon is actually closer to the Sun than the Earth is. The Moon looks like a waxing crescent because we're seeing only part of the lit side of the Moon.

It's important to realize that when and where we see the Moon in our sky is related to the phases of the Moon. Think about the new Moon for a second. When we look at the new Moon in the sky, it's in the same direction as the Sun. If it's in the same direction as the Sun, we won't see it at night because it's up when the Sun is up. But that also means it will rise with the Sun; it will be highest at noon when the Sun is highest; and then it will set with the Sun at sunset. The same is true for the other phases. Think about the full Moon up there in the sky. The full Moon is opposite the Sun in our sky. That means that as the Sun is setting, the full Moon is rising; or as the Sun is rising, the full Moon is setting.

Now let's look at the first quarter Moon in the sky. The first quarter Moon in the sky will be highest in the sky at sunset. It will rise six hours earlier at noon, and it will set six hours later at midnight. The third quarter Moon always is highest at sunrise in the morning. It rises up at midnight and it sets at noon. We often teach children that they way to tell the difference between a waxing and a waning Moon is that a waxing Moon is lit up on the right side, where a waning Moon is lit up on the left side. The problem with doing this is that it's opposite in the Southern Hemisphere. In the Southern Hemisphere, the waxing Moon is lit on the left, whereas the waning Moon is lit on the right. A better way to teach this is that the waxing Moon is always visible in the evening. Going back to our diagram you will see that in going from new Moon to full Moon, those phases of the Moon are always up between noon and midnight. That is, they're always up in the evening sky. The waning phases are always up in the morning; so the way to tell the difference between a waxing and a waning Moon is if you go outside and see a crescent Moon in the sky at 9:00 o'clock in the evening, it must be a waxing crescent Moon.

I think a project that everyone should do at least once in their lives is go out and observe the Moon every night for a month. Start with a new Moon or a waxing crescent Moon in the early evening. Draw the shape of the Moon and note its location in the sky, especially relative to where the Sun set. Every day thereafter, go out at about the same time, draw the Moon and how it appears in the sky, and do this for a full cycle of phases. Remember that after the full Moon, during the waning phases, you may have to be up in the early morning hours to see the Moon.

The unlit portion of the Moon that we see often appears to glow, you'll often see that the unlit portion is lit up. The unlit portion is lit up because this is sunlight reflecting off the Earth. It's called "earthshine" or "ashen light," or sometimes the very poetic, "the old Moon in the new Moon's arms." The reason that we see this unlit portion lit up by the reflected light from the Earth is, imagine what would happen if we had somebody standing on the surface of the Moon looking back at the Earth. Somebody standing on the Moon, looking back at the Earth would see a Gibbous Earth. They would see an almost-full Earth in the sky, which means the Earth will be quite bright, and all that light shining on the Moon is what lights up that unlit portion in the sky.

The Moon always presents the same face to the Earth. We always see that one face, and this is something that you can even see with the naked eye. The dark side of the Moon is never visible from Earth. It's been photographed by spacecraft and we see that the far side of the Moon looks very different than the near side of the Moon does. Another common misconception is that if the Moon always keeps the same face towards us, that the Moon isn't rotating. In fact, in order to keep the same face towards us, the Moon does in fact have to rotate.

Imagine we have the Earth sitting in the middle with the Moon orbiting around it. Imagine that we draw an astronaut or a mountain on the Moon. If the Moon is not rotating, that astronaut is always facing in the same direction.

So, as the Moon goes around the Earth, you can see that the Moon is not rotating because our little figure on the Moon is always pointing off the top of the screen. But now look at what people on Earth can see. People on Earth can see the side with the figure and then two weeks later, when the Moon is on the opposite side of the Earth, they can see the other side of the Moon. If the Moon doesn't rotate, we'd get to see all sides.

If we want to keep that figure, that astronaut, always facing towards the Earth, the Moon has to rotate. As it revolves around the Earth, it has to rotate a little bit to keep that astronaut pointing towards the Earth. Our Moon rotates on its axis in the same length of time it takes to revolve around the Earth. That is, it orbits the Earth once a month and it spins on its axis once a month. This is has arisen because of the tides, not only the tides that the Moon raises on the Earth, but the tides that the Earth raises on the Moon.

The dark side of the Moon, I think should properly be called the "far side of the Moon" because the dark side of the Moon isn't really dark. It gets just as much light as the near side. For example, if you think about what happens during a new Moon, the unlit portion of the Moon is facing towards the Earth but the lit portion of the Moon, that side that's facing away from the Earth is fully illuminated. They're getting sunlight on that far side of the Moon. So it's not dark, it's just our knowledge of the far side was dark, which is how it got the name "dark side of the Moon."

You might expect that if the Moon always keeps the same face pointed toward Earth, that we'd only see half of the Moon. However, the Moon appears to rock back and forth in the sky; and the reason for this is the Moon's elliptical orbit around the Earth. Because its rotation can't compensate for that very quickly, sometimes we get to peek a little bit around the right side and sometimes a peek around the left side. And because the Moon's orbit is tilted with respect to the ecliptic, sometimes we get to see a little above the Moon, and sometimes a little below the Moon. Altogether about 59 percent of the Moon is visible to us on Earth, as we can see in this animation of the rocking back and forth of the Moon.

The Moon is the first object that many people view through a telescope, and they're surprised to see a world that has mountains, craters, and lava flows dotted across the surface. After Saturn, the Moon happens to be my favorite object to look at through a telescope. It's an interesting world because today the Moon is geologically dead. Its small size means that it cooled off long ago and its lack of atmosphere means that there's no erosion today. In fact, its surface is a fossil of the early solar system. A lot of what we've learned about the formation of the planets and the conditions in the early solar system came from studying the surface of the Moon because it has preserved what our solar system was like.

The best time to observe the Moon through a telescope is around first or third quarter. You might think that the full Moon is the best time since the whole side of the Moon facing us is lit up and you could see everything on the Moon. But let me show you two images of the Moon and compare these two images to see why first and third quarters are better. This image is a composite of a first quarter Moon and a third quarter Moon and has been seamlessly stitched together down the center to produce a picture that shows us the full face of the Moon.

You'll notice that there are a lot of craters down on the southern part of the Moon, the bottom part of the Moon, and lots of dark, flat areas called the "*maria*" or the "seas." How can we tell that this is two photographs of the Moon stitched together? Look at the light on the craters on the Moon. On the right-hand side of the image, we see a crater that's highlighted with the arrow, and on that crater the light is coming from the right. We know that because the crater is bright on the left and dark on the right-hand side, which means the light is coming from over on the right-hand side. For a similar crater, on the left half of the image it's lit on the right and dark on the left, which means in that case the light is coming from the left-hand side.

Of course this can never happen with the real Moon. The light can't be coming from two directions at once, so we've put these two photos together. Let me show you an actual image of the full Moon, and in this case now, the illumination is coming down from straight above. There are no deep dark shadows on the Moon and without those shadows, we can't see all the craters on the surface as well as we did before. Some of the craters stand out, like Copernicus and Tycho are much more visible because of their rays' systems during the full Moon, but all those other craters disappear.

The best place to observe the Moon is along the day/night line, called the "terminator." And it's along the terminator where the shadows are darkest and deepest and they allow you to see the most features on the surface of the Moon. Night after night, as the Moon goes through its phases, the terminator will move across the surface of the Moon, highlighting different parts of the Moon.

When we look at these images, we can also see that the surface of the Moon consists of two different types of terrain. There are those lighter-colored, heavily-cratered highlands and the dark *maria*, which are very smooth. These two regions on the surface of the Moon have different ages, just like pieces of the Earth have different ages. So if you look at the highlands on the Moon and you look at the *maria* you might ask the question, which is older? The answer, surprisingly, is that the *maria* are the younger and the highlands are the older. You might have been thinking of erosion if you thought that the *maria* were the older ones because they've been worn down over time. But in fact, what's going on is that there is no erosion on the surface of the Moon due to the lack of wind and rain and ice on the Moon. Once you do something on the Moon, once you make a crater or a lava flow or a mountain, there's nothing to wear it down.

So how old are they precisely? The highlands are between 4.1 and 4.4 billions years old, whereas the *maria* are between 3.0 and 3.8 billion years old. The *maria* are ancient lava flows. Their name is Latin for "sea" and it comes because early observers noticed these smooth regions and thought they were filled with water. The *maria* cover about 31 percent of the near side of the Moon.

You might have noticed in the images that many of the *maria* are round in shape. This is because the lava flows filled in low lying areas caused by giant impacts on the surface of the Moon. The Moon is covered in craters of all sizes. Essentially all the craters on the Moon are the result of impacts from meteoroids or asteroids or comets striking the surface of the Moon. A good rule of thumb is that the size of the crater is about 10 times the size of the body that made it. If we take a crater like Copernicus, which is 58 miles or 93 km across, it was made by an object that was probably about 5 to 6 miles, or 9 to 10 km, in diameter.

The rays that you see around many large, fresh craters are material there were ejected by an impact. Large craters, like Copernicus pictured here, have a central peak and they have materials spread out around them that were spread out by the impact. Also, you can see in the walls of Copernicus terraces where some of the material has slumped down over time.

On the Moon, "young" is a relative term. These young, fresh craters, like Copernicus, are actually hundreds of millions of years old. They only look young and fresh because of the lack of erosion and weathering such as what we have here on Earth. You might ask, why is the Earth not covered in craters? In fact, it was a long time ago. In fact, because it's larger than the Moon, the Earth has been hit far more often than the Moon has. On the Earth, over time, wind, rain, ice, have worn down these craters. Only young craters, like Great Meteor Crater in Arizona, look young and fresh to us today.

Let me show you a crater on the Earth and one on the Moon so that you can see the difference between the two. Both of them are the same size, about 60 miles in diameter. On the left, we have the crater Copernicus on the Moon. Notice how rugged that crater is and how fresh all of its features look. On the right, we have a similar size crater on Earth. But in this case, thanks to wind and rain and erosion over time, it has been completely worn down so that all that's left is a lake filled with water. Interestingly, the crater on the Moon is 800 million years old while the crater on Earth is only 200 million years old. The lunar crater is four times older, and yet it's much better preserved. This is why geologists are fascinated by the Moon. It is, literally, a time capsule that has laid out for us the early history of the solar system.

Now that we've seen the Moon, let's move on to talking about the Sun. As we learned in Lecture 2, the Sun appears to move through the sky on a path that astronomers call the "ecliptic" as the Earth revolves around the Sun. As we said then, every year it passes through a series of 12 constellations called the "zodiac." It also happens to pass through a 13th constellation, Ophiuchus, that isn't part of the classical zodiac.

From the Earth, we see this as a shifting of the position of the Sun in the sky. In January, we see the Sun superimposed on the stars of Sagittarius. By February, the Sun has shifted in the sky because of the Earth's motion around the Sun, so that it's in front of the stars of Capricornus. A month later, in March, it is against the stars of Aquarius. Since the Earth orbits the Sun in a year, and it takes 365 days for the Sun to complete one circuit of the sky, which is 360 degrees, the Sun appears to move about one degree, day by day against the background stars.

On a celestial sphere, you can see that the Sun's path across the sky is tilted by about 23.5 degrees with respect to the Earth's equator projected against the sky. That is, the Sun can be above or below the equator depending on the time of the year. As the Earth orbits around the Sun we see marked on the Earth the North and South Poles. The North Pole is always pointing at the bright star Polaris in the sky. It's always pointing at the North Celestial Pole. At various times of the year the North Pole is tilted towards the Sun and it's tiled away from the Sun. Now, let's project the Earth's equator out into space. This plane represents the Earth's equator. Now look at the Sun. You will see there are times that the Sun is above the equator and times that the Sun is below the equator. On March 20, the Sun is passing from below the equator to above the equator. Then, on six months later, it's passing from above the equator to below the equator. On June 21, the Earth is tiled towards the Sun, whereas on December 21, the Earth is tilted away from the Sun.

This 23.5 degree tilt of the Sun is what causes the seasons on Earth. It's a really common misconception that the seasons are due to our distance from the Sun. We're actually closest to the Sun around January 3rd of every year, and we're farthest from the Sun around July 4th of every year. In the Northern Hemisphere, it's hottest in July when we are farthest from the Sun and it's coldest in January when we are closest to the Sun. The tilt of the Earth is responsible for the seasons, because when a given hemisphere is tilted towards the Sun, the days are longer and the Sun reaches a higher point in the sky. The higher the Sun is, the more efficient it is at heating the Earth. At mid-northern latitudes, the amount of energy received from the Sun is more than twice as large at the summer solstice than at the winter solstice, and the days are 1.5 times longer.

There are four important marker days that we have to know during the course of a year. These are the vernal equinox, or spring equinox, around March 20th when the Sun crosses the equator going from south of the equator to north of the equator. Three months later we reach the summer solstice on or about June 21st, and that's when the Sun is farthest north in the sky. The autumnal, or fall equinox, occurs on about September 22nd, and that's when the Sun crosses the equator going from north of the equator to south of the equator. The winter solstice occurs on or around December 21st. It's when the Sun is at its farthest south.

Another common misconception is that the Sun always rises due east. Only on the vernal and autumnal equinoxes when the Sun is on the equator does it rise due east and due west. Look at this diagram, again for an observer at mid-northern latitudes. From March 21st until September 21st, when the Sun is north of the equator, it actually rises in the northeast, makes this giant path across the sky and sets in the northwest. On March 21st and September 21st, when it's on the equator, it rises due east, makes an arc through the sky, and sets due west. You can see that that circle looks like exactly half of a circle. In fact, that's where the "equinox" comes from. It means equal night, and the days and nights are equally long. From September 21st until March 21st, the Sun is south of the equator. It rises in the southeast. It sets in the southwest. It doesn't get very high, and as you can see from the small arc, the Sun is not up in the sky for very long. The changing location of the rising or setting Sun is something that everyone should make an effort to note. Take a look at this image for example. This is an image taken on three days of the year. On the summer solstice, on the equinoxes, and on the winter solstice. You can see the rising position of the Sun changing in the sky.

Tracking the location of the rising Sun is probably humanities' earliest calendar. Huge stone monuments, such as Stonehenge, were built in an effort to track the annual rising and setting of the Sun so that these ancient peoples could track the passage of the seasons. At Stonehenge for example, there is a distant heelstone. On the Summer solstice the Sun rises right over the heelstone and on that day only. Every other day of the year it rises somewhere else along the horizon. They could use this heelstone for marking the passage of the seasons.

The motion of the Sun north and south of the equator is responsible for the changing length of our day. When the Sun is north of the equator here in the Northern Hemisphere, the days are longer because the Sun is making that large path in the sky. In the winter time when the Sun is south of the equator, it makes that small path through the sky, and that small path means that the Sun isn't up for very long.

Since the Sun can be, at most, 23.5 degrees north or south of the equator, the only people that can ever see the Sun straight overhead are those whose latitude is within 23.5 degrees of the equator. It's a common misconception that the Sun is straight overhead at noon on every day, or even straight overhead at noon on the summer solstice. In fact, unless you live between 23.5 degrees south latitude and 23.5 degrees north latitude, the Sun will never be straight overhead. The Tropic of Cancer, in fact, at 23.5 degrees north latitude, marks the location on Earth where the Sun is straight overhead on June 21st, the summer solstice. If you live farther north than the Tropic of Cancer, the Sun will never be straight overhead. The Tropic of Capricorn, at 23.5 degrees south, marks the location on Earth where the Sun is straight overhead on the December 21st, the winter solstice. If you live farther south than the Tropic of Capricorn, the Sun will never be straight overhead for you either. For people living between the tropics, the Sun will be straight overhead twice a year. Once as the Sun is heading north and again as the Sun is heading south. Even at the equator, the Sun is only straight overhead at noon on the equinoxes, twice a year.

Now imagine what happens near the poles on the solstices. On December 21st, on the winter solstice, the Sun is 23.5 degrees below the equator. This means that the Sun doesn't rise for anyone within 23.5 degrees of the North Pole, called the "Arctic Circle." For people within 23.5 degrees of the South Pole, the Sun never sets. That is, as the Earth turns around, they're in continuous daylight, and that's inside the Antarctic Circle. On June 21st, the summer solstice, the situation is reversed. For those up at the North Pole, inside the Arctic Circle, the rotation of the Earth carries them round and round so that they're always in daylight. They get 24 hours of continuous daytime. While those in the Antarctic Circle get 24 hours of continuous nighttime. This gives rise to the idea of the "midnight sun." As shown in this photograph taken on the summer solstice within the Arctic Circle, the Sun never sets but rather goes round and round the sky, and even at midnight, the Sun is above the horizon, low but still above the horizon.

Let's finally turn to observing the Sun. It's important to know that you should never look at the Sun without eye protection. Even a momentary view of the Sun through unprotected binoculars or a telescope can cause severe and permanent damage or blindness. You can safely view the Sun without a telescope using special-built solar glasses or by viewing the Sun through number 14 welder's glass, or by projecting the image. There's an old standby, where you take a piece of paper and you poke a hole in the piece of paper and project it down on another piece of paper. And the problem with doings this that I found is that the image of the Sun is very small and it's very faint and it's hard to see. There are much better ways to project an image of the Sun down on a piece of paper so that you can see it.

I have here a couple of pieces of equipment that would allow us to do this. The sunspotter is a small telescope that has a glass lens that focuses the light down onto a piece of paper and it makes a nice large image of the Sun. On this image of the Sun you can see sunspots and other features on the Sun. Another example is the solar scope, which does the same thing. It has a small telescope that projects an image of the Sun on the inside of the box. Both of these are very safe for using because you're not putting your eye up to the telescope. There's no chance that the light from the Sun would be focused on your eye. In both of these cases, the image of the Sun is projected safely down on a piece of paper.

Without a telescope, it's possible to see sunspots on the Sun but only when they're really large. With these telescopes you can see sunspots on a regular basis. Sunspots are magnetic storms on the surface of the Sun and they appear dark because they're cooler than the rest of the surface, though they're actually still brighter than the full Moon in the sky. The Sun goes through an 11-year cycle called the "solar cycle," when we can see many spots on the Sun, and at times we see no spots on the Sun. Every 11 years it goes from being almost spot-free to having many spots. Observing sunspots over the course of a few days will allow you to track the rotation of the Sun. At the equator it takes the Sun about 25 days to rotate once.

You can also view the Sun through a telescope. There are a couple of different ways to do this. One way to view the Sun safely through a telescope is to use a special-built solar filter. A special-built solar filter like this you can put on the front of your telescope and it will block all the damaging ultraviolet and infrared light from the Sun and let through only the visible light from the Sun. A filter like this, you have to be very careful to make sure that when it goes on the front of the telescope that there's no chance that it will fall off. I like to take the extra precaution of actually using a piece of tape to hold this onto the telescope when I put this filter on my telescope.

One thing you should be aware of is old telescopes where the solar filter is actually down by the eyepiece. When the solar filter is down at the eyepiece and there's nothing covering the front of the telescope, all the light of the Sun is focused on that one little filter, and that filter can get very, very hot and overheat. If it overheats and cracks, the unfiltered light from the Sun will come right into your eye.

The last piece of equipment I'd like to talk about is a special kind of solar telescope that looks at only a particular color of light coming from the Sun. These are called "H-alpha telescopes" and they look at one particular red color of light that comes from hydrogen atoms. This is a special-built solar telescope. You can't use it for observing objects in the night sky. But what it does allow you to see are the beautiful prominences on the limb of the Sun sticking up above the limb. You can buy a special-made solar telescope like this. You can also buy H-alpha filters that will sit on the front of your telescope and let through only that one particular color of red light.

The path of the Moon as it orbits the Earth is tilted by about five degrees with respect to the Earth's orbit around the Sun. The Sun's annual path through the sky is also the path of the planets in our sky. And in the next lecture we'll talk about using a telescope to see the planets in detail.

# Observing the Planets with a Telescope
## Lecture 5

Lecture 5: Observing the Planets with a Telescope

> **Because of Jupiter's rapid rotation, new features are constantly rotating into view on the planet. In fact, since it rotates in 9.8 hours, in a long winter night, it's possible to observe the full surface of Jupiter.**

Our solar system has three kinds of planets: **terrestrial planets** in the inner solar system (Mercury, Venus, Earth, and Mars), **Jovian planets** in the middle solar system (Jupiter, Saturn, Uranus, and Neptune), and cold, icy bodies in the **Kuiper belt** at the outer edge of the solar system (Pluto and Aris).

Because Mercury and Venus orbit closer to the Sun than the Earth does, both planets go through a full set of phases, like the Moon. Typically, these planets are not visible when they are on the far side of the Sun. Mercury is the smallest of the terrestrial planets and a speedy one; it orbits the Sun every 88 days. Like our Moon, Mercury is covered in craters and giant impact basins. Viewing Mercury is challenging because it is small and it is never very far from the Sun. Venus is a near twin to Earth in size, but it has a stifling atmosphere made mostly of carbon dioxide. The average surface temperature on Venus is around 870° F due to a strong greenhouse effect

© Digital Vision/Thinkstock.

**Venus, one of the rocky, terrestrial planets of the inner solar system.**

from this thick atmosphere. Venus appears larger than any other planet and is the third brightest object in the sky, after the Sun and the Moon.

Mars is about half the size of the Earth, which means that it's difficult to see features on its surface. In order to see details on this planet through a telescope, it is essential to observe Mars at opposition, when it is closest to the Earth, which occurs only every 26 months. Mars displays a number of interesting features, such as polar caps, dust storms, and clouds. The surface of Mars is also covered with high-contrast light and dark terrains.

**Saturn, a Jovian planet, whose water-ice rings span more than 270,000 kilometers.**

Jupiter, Saturn, Uranus, and Neptune are made mostly of hydrogen and helium, the same elements that make up the Sun. When you look at these planets through a telescope, you don't see their surfaces but the upper layers of their thick atmospheres. Jupiter is the largest planets in our solar system and one of the best objects to view. You can easily see bands of clouds on the planet and the four giant Galilean moons. The Great Red Spot, a rotating, high-pressure storm on the southern edge of the south equatorial belt, is probably the most famous feature on Jupiter. Saturn is another spectacular object to view through a telescope. It comes to opposition once every 54 weeks. The rings of Saturn, made of countless orbiting particles of water ice of all sizes, are the planet's most well-known feature. Like Jupiter, Saturn has belts and zones (although they are harder to see on Saturn) and more than 60 moons.

Uranus was the first planet to be discovered. It is faint and has a slow orbit and, even through a telescope, appears as a featureless blue-green ball. Like Uranus, Neptune can be seen moving through the sky, but it is very small and appears as a featureless blue disk. Out at the edge of the solar system is the Kuiper belt, a region of cold, icy bodies that includes Pluto. Even in a large telescope, Pluto appears as nothing more than a very dim star at the edge of visibility. A number of asteroids are also visible through a telescope or a pair of binoculars. Most of these are small, rocky bodies orbiting between Mars and Jupiter. ■

## Important Terms

**Jovian planets**: Giant gaseous planets in the middle solar system: Jupiter, Saturn, Uranus, and Neptune.

**Kuiper belt**: A region of cold, icy bodies at the edge of our solar system.

**terrestrial planets**: Rocky planets in the inner solar system: Mercury, Venus, Earth, and Mars.

## Suggested Reading

Dickinson, *Nightwatch*, chap. 7.

Ridpath, *Norton's Star Atlas*, chap. 3, 65–86.

## Questions to Consider

1.  When will Venus next be at its greatest elongation?

2.  How does the phase of Venus change around the time of greatest elongation? Use the diagram in the lecture to determine how it will change and, if you can, observe Venus through a telescope once a week for 4–6 weeks before and after greatest elongation. How does the phase of Venus change during this time?

3.  When will Mars again be closest to Earth? How far will Mars be from Earth at its closest compared to average? How large will it appear in the sky?

4.  How long does it take Jupiter's moons to orbit the planet? Use a telescope to see the moons of Jupiter. Every day over the course of two weeks make a quick drawing of the locations of the Moons relative to the planet and watch as they orbit the planet. This is exactly what Galileo did over 400 years ago and led to one of his most important discoveries.

5. How fast does Jupiter rotate? You can use the Great Red Spot on Jupiter to watch the planet rotate. Use a table from the Internet or an astronomy magazine to determine when the Great Red Spot on Jupiter will next be visible and when it will again be visible after that. If you have a telescope, observe the planet. Can you see the Great Red Spot?

6. How will Saturn's rings appear this year? The rings appear edge on twice during every 29.5 years as Saturn orbits the Sun. If they were edge on in September 2009, how will they appear tonight? Will they be wide open or edge on?

# Observing the Planets with a Telescope
## Lecture 5—Transcript

In the last lecture, I described how the Moon goes through its phases. Many people are surprised to find out that both Mercury and Venus go through phases as well. In this lecture, I'll describe how best to see each of the planets through a telescope. I'll tell you what to look for and what you should look for when you're looking at the planets through a telescope. When I bring a telescope to a school or other public viewing event, the first thing that most people want to see is any bright planet that happens to be visible in the sky at that time. But to see much detail in the planets will require a telescope and it will require moderate to high magnifications, moderate magnifications being 50 to 100 times or high magnification being 100 times or more.

Before we talk about the planets as they will appear through a telescope, we need to know a little bit about the planets in our solar system. I think it's much more important to know that there are three kinds of planets in our solar system rather than the actual number of planets. In the inner solar system we have the hard rocky planets: Mercury, Venus, Earth, and Mars. These are the terrestrial planets. They're made of rock just like the Earth is. You could don a spacesuit and you could go on any one of these planets and you could walk around on the solid surface. It wouldn't be surprising that you would see mountains and volcanoes and craters and all the other features we see on planets like the Earth and the Moon.

In the middle part of the solar system we have Jupiter, Saturn, Uranus, and Neptune. These are the gas giant planets, or the jovian planets. If you were to take a parachute and parachute down through the atmosphere of one of these giant planets, you would continue to sink deeper and deeper into the planet. They're completely gas throughout. They have no solid surface. At the outer edge of the solar system we have the cold, icy bodies. There's a belt of these cold, icy bodies called the "Kuiper Belt." It contains probably millions of objects. The two largest ones that we know of right now are Pluto and Aris.

In this lecture, I will include some images from the Hubble Space Telescope and other spacecraft that have gone and viewed the planets up close. I just want to warn you that you won't see that much detail when you look through a telescope at the planets. But these images are important because they help us to understand what it is that we're seeing and what we should look for through a telescope. I think the natural order of arrangements for planets in the solar system is to discuss the planets in order from Mercury outward. So let's start in the inner solar system and start with the planet Mercury.

Because Mercury and Venus orbit closer to the Sun than the Earth does, both planets go through a full set of phases like the Moon. Here's a diagram that shows the orbit of one of these inner planets. When it's on the far side of the Sun, we're looking at the fully- illuminated phase and so we see it in the full phase. Typically the planet is not visible when it's on the far side of the Sun because it's up close to the Sun in the sky, and so we don't often see Mercury or Venus at full phase. As the planet orbits around the Sun, it goes through a Gibbous phase and then a quarter phase, and then as it comes between the Earth and the Sun, we see a thin crescent when most of the lit side is facing away from us. And finally as the planet passes between the Earth and the Sun we see the new phase of the planet.

I love this picture of Venus, or these pictures of Venus I should say. It's a series of pictures that were taken over a few months as Venus went through its phases as it orbited the Sun. I've flipped the image so that you could see what a full orbit of Venus would look like. A couple of things to notice. You can see it going through its phases. At the top it starts at the Gibbous phase and then comes down through the crescent. Of course, the new phase is invisible, there's not much to see, and the full phase on the far side of the Sun isn't visible either. But the other thing to notice about Venus is its change in size. When it's on the far side of the Sun it's much smaller than when it's between the Earth and the Sun and much closer to us, so not only will you see Venus going through its phases, you'll see its size changing as well.

The phases of Venus were one of Galileo's crucial observations that were determined that the Sun was the center of the solar system and not the Earth. Originally, if the Earth was the center of the solar system and Venus orbited around the Earth, it wouldn't go through phases like this. So one of the proofs that the Sun was at the center of the solar system was Galileo's observation that Venus went through phases.

Let's turn our attention now to Mercury. It's the smallest planet of the terrestrial planets. It's only 38 percent the diameter of the Earth, and it's only 5.5 percent the mass of the Earth. It's a speedy planet. It orbits the Sun every 88 days at an average distance of 0.4 astronomical units, or only 40 percent the distance from the Earth to the Sun. Spacecraft have shown us that Mercury is very much like our Moon. It has no atmosphere so there's no erosion to wear down its features; and it's a battered planet. It's covered in craters and giant impact basins. But it's so small and we generally see it from far enough away that none of these surface features are visible in telescopes from Earth.

At best with Mercury, you can watch it go through its phases just like Venus went through its phases just a second ago. But even this is a really challenging thing to do because Mercury is small and it's never very far from the Sun in the sky. In fact, when it's on the far side of the Sun from us, Mercury is only five arcseconds across. It gets as large as 13 arcseconds when it's between the Earth and the Sun. The best time for viewing Mercury is when it appears farthest from the Sun. You'll remember, that was called its greatest elongation and then it only appears seven arcseconds in size. Now, Mercury can be observed with a telescope during the daytime but you have to be exceptionally careful here because you never want to point your telescope at the Sun. If Mercury is close to the Sun in the sky, you have to be very, very careful.

At its greatest elongation, Mercury is never more than 28 degrees from the Sun. With an elongation of only 28 degrees, that means Mercury is always low to the horizon right after sunset or low to the horizon right before sunrise. These low altitudes mean the trees and buildings can block your view of Mercury and atmospheric turbulence will be at its worst, so you won't get a good clear view of the planet. Because Mercury orbits the Sun in 88 days, it's only visible for a few days before and after the greatest elongation. It's orbiting so quickly that after elongation, a few days later, it's already lost in the bright light of twilight. Even though Mercury can reach an apparent magnitude of −0.7, it's probably the least observed of the naked eye planets.

You might think with Mercury being the closest planet to the Sun that it's also the hottest planet in our solar system, but that's not correct. Venus is the hottest planet in our solar system even though it's almost twice as far from the Sun as Mercury is. Venus is a near twin to Earth. It's 95 percent the diameter of Earth and 82 percent the mass of Earth but the similarities end there. Venus has a stifling atmosphere made mostly of carbon dioxide. On the surface, the atmospheric pressure is 92 times the atmospheric pressure on Earth, and this thick atmosphere acts like a blanket to trap in the heat. Although Venus is farther from the Sun it's hotter than Mercury is. The average surface temperature on Venus is around 870 degrees Fahrenheit, around 465 degrees Celsius, due to a strong greenhouse effect from this thick carbon dioxide atmosphere. Venus orbits the Sun every 225 days at an average distance of 0.72 astronomical units and that means Venus gets closer to the Earth than any other planet. When it's on the far side of the Sun from us, it's as small as 10 arcseconds, and when it's between the Earth and the Sun it's as large as 65 arcseconds or just over an arc minute in size. That means Venus appears larger than any other planet. At it's brightest it can reach magnitude −4.7, which makes it the third brightest object in our sky after the Sun and the Moon.

Although Venus appears brighter and larger than any other planet, we can't see any features on the surface of Venus through a telescope, and that's because the planet is completely enshrouded in a thick layer of permanent high-altitude clouds. These clouds are made mostly of little droplets of sulfuric acid and they're suspended between 18 and 43 miles, about 30 to 70 kilometers, above the surface. They're excellent reflectors of light, which explains why Venus appears so bright in our sky, but they don't allow us to see the surface. Venus reaches greatest elongations of 45 to 47 degrees from the Sun, and at the time of greatest elongation, bright Venus is visible for many weeks in the western sky just after sunset or in the eastern sky just before sunrise. At its greatest elongations Venus appears about half phase and it's about 24 arcseconds in diameter. With a good pair of binoculars or a telescope, you can watch Venus go through its phases.

Mars is the next planet out that we're interested in studying. Mars is one of the most famous planets for looking at through a telescope, but I must also warn you, it's a very difficult planet to see features on the surface because it's small. It's only 53 percent the diameter of the Earth, half the size of the Earth, and it's 11 percent the mass of the Earth. It orbits the Sun every 687 days at an average distance of 1.5 astronomical units . In order to see details on Mars through a telescope, it's essential to observe Mars at opposition when it's closest to the Earth. Opposition with Mars only occurs every 26 months, or every other year, so you don't get an annual opportunity to see Mars at its best. Mars orbits the Sun once every 1.88 years in a very elliptical orbit. There are times in its orbit when it's close to the Sun and times in its orbit when it's far from the Sun. Because of this there are times when Mars and Earth approach close to one another and Mars is particularly close to the Earth—this is called a "perihelic opposition"—and times when Mars is far from the Sun at the same time that Earth passes it and Mars happens to be particularly farther than average during one of these close passages. And that's an "aphelic opposition." During a perihelic opposition it can come as close as 56 million kilometer, while in the worst case at the aphelic opposition it's no closer than 101 million kilometers.

Mars appears significantly larger during a perihelic opposition than an aphelic opposition. During one of these perihelic oppositions, Mars can be as big as 25 arcseconds across, and it never appears any larger than that. At its brightest, it will appear as a magnitude −2.9 bright light in the sky looking very much like a star. During aphelic oppositions, during these close approaches when Mars just happens to be farther from the Sun than average, it gets no larger than 14 arcseconds in size, and it appears at magnitude −1.0. When it's on the far side of the Sun from the Earth, Mars is tiny, only 4 arcseconds across. Mars appears good around these times of opposition, and it looks very good through a telescope about a month before opposition and about a month after opposition. Outside that window, the planet appears very small and the surface features are very difficult to see.

Here's a table that shows you oppositions of Mars and the distance from Earth to Mars during these oppositions so you can plan out when a good time to observe Mars would be and how good it might appear in our night sky. Mars displays a wealth of interesting and changing features to a dedicated observer. You can see on Mars polar caps, dark markings on the surface, dust storms, clouds, and hazes. Like Earth, Mars has a north and a south polar icecap. On Mars they're made mostly of water ice with frozen carbon dioxide. The brilliant white polar caps on Mars grow and shrink during the Martian year just like the polar caps on Earth grow and shrink with our seasons. The growing and shrinking of these polar caps is one of the easiest and most-interesting things to observe on Mars. From one opposition to the next opposition you'll catch Mars in its different seasons and the polar caps will look distinctively different. But even during a single opposition, for those few months when Mars looks good in the sky, you may notice weekly changes in the polar caps either shrinking or growing in size.

At perihelic oppositions, when Mars is particularly good, the southern polar cap is tilted towards the Earth. The opposite is true on the poorer oppositions when Mars is farther from us. In that case it's the northern polar cap that's tiled towards Earth, so we get the better view of Mars' southern polar cap from here on Earth. Even in the Martian summer, the polar icecaps don't completely disappear. These permanent icecaps are believed to be mostly water ice with a little bit of frozen carbon dioxide in them, but the growing and shrinking icecaps, these seasonal icecaps, are mostly frozen carbon dioxide. The polar caps, especially in the autumn and the winter, can be covered with clouds called "polar hoods," and these polar hoods are visible through an eyepiece. They appear as an indistinct bluish-white haze up on the polar caps of Mars.

The surface of Mars is covered with high-contrast light and dark terrains. These markings are visible in telescopes as small as 60 millimeters. The dark markings were given names taken from terrestrial geology or mythology by the Italian astronomer Giovanni Schiaparelli. Most of the markings represent changes in albedo, or reflectivity or brightness of the surface. They don't always correspond to a specific feature such as a volcano, a canyon, or a giant impact crater, though some of them do. Mars rotates once on its axis every 24 hours and 40 minutes. If you go out and observe Mars this evening at 9:00 p.m. and go out tomorrow night at 9:00 p.m., you'll see many of the same features on Mars because the Earth has completed one rotation and Mars has completed one rotation. But because they don't exactly match, Mars takes 24 hours and 40 minutes to rotate, the features on Mars will be shifted by just a little bit. In order to see the whole surface of Mars you have to observe Mars at various times: early evening, middle of the night, early morning. And if you do that for a few weeks, you'll eventually get to see the whole surface of Mars.

Over time these dark markings are known to change. They get brighter and they get darker mostly due to dust flowing across the surface of Mars. It's interesting from one opposition to the next to see how these various familiar features have changed in appearance. The famous Martian canals of Percival Lowell and Giovanni Schiaparelli are an optical illusion and you won't see these through a telescope. Of course close-up photographs of Mars taken by spacecraft show us that the surface of Mars is not carved into an intricate series of canals. In addition to the polar hood clouds, Mars frequently has other clouds that might be visible. Some clouds form over the giant volcano Olympus Mons and the volcanoes on the Tharsis Rise. Clouds or frost can be seen as a white haze in the morning or early evening on the terminator, the day/night line. So when you look at Mars, look on both sides of the planet where the Sun is just rising and the Sun is just setting to look for this frost or this haze.

The Martian atmosphere is very, very thin and it's primarily made of carbon dioxide like the atmosphere on Venus. The atmospheric pressure on Mars is only 0.7 percent the atmospheric pressure on Earth. To tell you how thin this is, it's equivalent to the pressure in the Earth's atmosphere at an altitude of 110,000 feet. Even though the atmosphere is thin, the atmosphere can pick up little small grains of dust and blow them across the surface, and these winds can develop into huge dust storms which appear as yellow, hazy patches on Mars. During perihelic oppositions, when Mars is closest to the Sun and the surface heating is greatest, these dust storms can develop into giant global dust storms that cover most or all of the planet. If you happen to observe Mars through a telescope during one of these global dust storms, the dust will blot out all the surface features. You won't be able to see any of the dark markings on the surface and Mars will appear as a featureless yellow ball. Observations over a few consecutive nights will allow you to track the development of these storms as they move across the surface, and at these perihelic oppositions, may develop into a global dust storm covering the surface.

Mars has two tiny moons. The first is Phobos, which stands for fear, and the other is Delimos, which is dread or panic, which is appropriate of course for the God of War, Mars. Both moons are likely captured asteroids as they're only 22 kilometers and 12 kilometers in diameter, but the moons are very difficult to see. First of all, they're faint at magnitudes 10.5 and 11.5, but they also happen to be very, very close to Mars, and the brilliant glare from Mars makes it difficult to see the tiny moons up close. To see the moons requires a moderate to large telescope, and you can either place brilliant Mars outside the field of view or put a little bar across the eyepiece, like a slip of paper or a piece of tinfoil, to block the glare from Mars so that you might see the moons peeking out on either side.

Mars is probably the best of the planets for trying to use color filters to make the features more noticeable. Blue and green filters make the polar caps, the clouds, and the hazes more visible, while red and yellow filters make the dark markings and the dust storms more visible. Most astronomical filters are color photographic filters and they use the Wratten photographic numbering system to distinguish the various colors. Common filters are the #15 deep yellow, #25 red, #58 green, and the #80A blue. And all these color astronomical filters come in both 1.25 and 2-inch sizes. So they'll thread right into your 1.25 and 2-inch eyepieces.

Jupiter, Saturn, Uranus, and Neptune are the gas giant planets. They're made mostly of hydrogen and helium, the same elements that make up the Sun. As I said earlier, these planets don't have a solid surface like the terrestrial planets, so when we look at them through a telescope we're not seeing their surfaces but we're seeing the upper layers of their thick atmospheres. Jupiter is the largest planet in our solar system. It's 11.2 times the diameter of the Earth and it's 318 times the mass of the Earth. In fact, the mass of Jupiter is greater than all the other planets put together. It orbits the Sun every 12 years at an average distance of 5.2 astronomical units . It's a little more than five times farther from the Sun as we are. At its maximum brightness, it's magnitude −2.9 making it the fourth brightest object in the sky after the Sun, the Moon, and Venus.

Jupiter is one of the best objects to look at through a telescope. Even though it's five times farther from the Sun than we are, about three and a half times farther away than Mars, because it's 11 times the size of the Earth, it makes it the planet with the largest average size. Venus can sometimes appear a little bit bigger, but on average, Jupiter is the largest. At opposition Jupiter ranges from 44 to 50 arcseconds across, nearly an arcminute in size. It's never smaller than 30 arcseconds across, even when it's on the far side of the Sun. Jupiter comes to opposition every 13 months and it's good to view through a telescope for a month or two before opposition and a month or two after opposition; so you have plenty of opportunities to see Jupiter.

When you look at Jupiter, you'll easily see bands of clouds on the planet and the four giant Galilean moons and possibly some features in those cloud bands. Jupiter's atmosphere is broken into a series of dark belts and bright zones by Jupiter's rapid rotation. Jupiter is 11 times the size of the Earth but it rotates in only 9.8 hours. The belts in the zones are upper layers of clouds and they consist mostly of ice crystals of ammonia, ammonium hydrosulfide, and water ice. Two main dark bands, two main dark belts I should say, dominate the disk of Jupiter, and they're called the "north and south equatorial belts." Because of Jupiter's rapid rotation, new features are constantly rotating into view on the planet. In fact, since it rotates in 9.8 hours, in a long winter night it's possible to observe the full surface of Jupiter.

Under exceptional observing conditions with high magnification and a large telescope you may actually see dozens of belts and zones on the planet. The bright zones are regions where warm gas is rising and condensing into bright clouds that are good reflectors of the Sun's light. The bright zones are higher in altitude than the dark belts. The dark belts are areas where cool gas is sinking back into the interior of Jupiter, and because this cool gas is rather transparent, it allows us to see deeper into the planet. And because we're seeing deeper into the planet these belts are darker in color. The belts range in color from dark gray to reddish brown, and the zones, the brighter ones, are usually white or yellow or ochre.

From year to year and decade to decade, the belts and zones are seen to change in appearance. They grow brighter and darker. They split into two. They combine with one another. They change colors. Jupiter is a dynamic planet that never looks quite the same from one opposition to the next.

The Great Red Spot is probably the most famous feature on Jupiter and it's a giant, rotating, high-pressure storm on the southern edge of the south equatorial belt. Remember, if your telescope flips the view upside down, the south equatorial belt may be the one that's on top when you look at it through an eyepiece. Although the storm appears to be a permanent feature, it has been seen for about 180 years and possibly as far back as 345 years, the size and color of the Great Red Spot are not constant. It changes from year to year. Keep in mind that when you look at this red spot through a telescope, it's about two to three times the diameter of the Earth. The Great Red Spot, although that's its name, actually changes in color. It can be a dark brick red, to a pale tan, or even white in color. Many smaller spots and ovals are also seen to appear and disappear in Jupiter's belts and zones. Some have lasted for decades as they grow and shrink and change colors and merge with one another and interact with the Great Red Spot. You can track these in a moderate-sized telescope. What causes all the colors in the Great Red Spot and on the belts of Jupiter is a great mystery right now. The three layers of clouds that have formed in these bands are all expected to be white in color. The color of this deep red color of the Great Red Spot, or the belts on Jupiter, may come from sulfur compounds in the atmosphere or even complex, organic molecules. But scientists just aren't sure right now.

At last count Jupiter had over 60 moons. Of these only four Io, Europa, Ganymede, and Callisto are easy to see in a telescope. The discovery of these four moons by Galileo ranks as one of the great astronomical discoveries of all time because at that time people were wondering, if the Earth was not at the center of the universe, if the Earth was really going around the Sun, how could the Moon keep up with the Earth? Galileo used the moons of Jupiter to show a moon could orbit a planet as the planet went around, and Earth wasn't sitting at the center. So like the phases of Venus, it was one of the key discoveries in overthrowing the geocentric or Earth-centered model of the universe.

Through a telescope or even a powerful pair of binoculars the moons look like four bright stars on either side of Jupiter. They orbit around Jupiter's equator, so they often appear in nearly a straight line as seen from Earth. From night to night as the moons orbit around Jupiter they shift positions. Io, the inner moon, completes a full orbit in 1.8 days, while Callisto requires 17 days to go around. The moons themselves are too small to show any surface features so mostly you just see these bright star-like objects moving around Jupiter. Since these moons orbit in Jupiter's equatorial plane, it's possible to see them transit in front of Jupiter and cast their shadows down onto the planet. The shadows appear as a small dark spot right on Jupiter, and this dark spot is easily visible through a telescope. Over the course of a few hours the dark spot moves across the planet. It is in fact a solar eclipse happening on Jupiter. Every six years, when the orbital planes of the moons lines up with the Earth's orbit around the Sun, we're seeing the moons edge on and it's possible for mutual events where one of the moons passes in front of the other moons and we see eclipses of the moons occulting or blocking one another.

Saturn, I think, is the most spectacular object to view through a telescope and it's my personal favorite. Saturn is 9.5 times the diameter of the Earth and it's 9.5 astronomical units from the Sun, so it's about twice as far away from the Sun as Jupiter is. Saturn comes to opposition once every 54 weeks. So you have one good opportunity every year to view Saturn. At opposition it's about 18 to 21 arcseconds in size and it can reach magnitude −0.5 during the best oppositions when the rings are wide opened, and it'll be a little bit dimmer during oppositions when the rings are edge on. We'll talk more about that in a second. Due to its large size and its great distance, Saturn is well placed for viewing for a few months before and a few months after opposition. You have a long good time to look at Saturn.

Of course the most famous feature of Saturn is its rings. The rings of Saturn are visible even in a small telescope or a really good high-power pair of binoculars. The bright rings of Saturn span more than 270,000 kilometers. The rings are exceptionally thin, only a few tens of meters thick. The rings of Saturn are not solid. They're made of countless particles of water ice ranging from the size of dust grains up to boulders. If you were standing inside the rings of Saturn, this is very much what you would see, particles of all sizes, maybe not as round as this, maybe much more jagged than this, but to give you an idea of how big they are, the typical dust grain is about the size of a person's head. Each of these particles individually orbits Saturn. It's important to know that Saturn's rings are not solid. In the densest part of the rings, the particles probably gently collide once every few hours.

The ring plane of Saturn is aligned with the equator of Saturn and it's tilted about 27 degrees with respect to Saturn's orbit around the Sun. At times, as Saturn goes around the Sun, we see the rings edge on and at other times we see the rings wide open. And in fact the rings appear edge on about every 15 years or so, and over the 30 year orbit of Saturn around the Sun we'll see the rings open up and then close and then open up again. There are three major rings that are visible through a telescope. The A ring is the outermost of the three. It's darker than the inner B ring, which is the brightest of the three rings, and just inside the B ring we have the very faint C, or Crepe, ring, which is partially transparent and it's very challenging to see. On a really good night you might notice between the A and the B rings a very thin, black gap between the rings and this is "Cassini's division." It's an actual gap in the [rings] that has been cleared out by one of Saturn's moons, Mimas.

Like Jupiter, Saturn has belts and zones. But the belts and zones on Saturn are much harder to see because these layers occur deeper down inside the planet and the overlaying layers of Saturn's atmosphere make it harder to see these belts and zones, so they're much harder to see. They're much more diffuse on Saturn than they are on Jupiter. Like Jupiter, Saturn has been seen to develop spots and ovals that change over a period of weeks or months. Like Jupiter, Saturn has many moons, over 60 at last count, most of which are small, captured asteroids or icy bodies. The easiest one to see is its large moon Titan because it's the second largest moon in the solar system and it's visible in good binoculars or a small telescope at magnitude 8.3. Of all the moons in the solar system, it's the only moon with a substantial atmosphere, about 1.5 times the surface pressure of Earth's atmosphere. A few other moons are visible in a small to moderate-sized telescope, including the tenth-magnitude moons Rhea, Tethys, and Dione, and possibly under good skies you might see twelfth-magnitude Enceladus. Other moons are much fainter and harder to see.

Next out in the solar system we come to Uranus. Uranus is a gas giant planet about four times the diameter of Earth. It takes 84 years to orbit the [Sun] at an average distance of 19.2 astronomical units. Uranus was the first planet to be discovered. When William Herschel was conducting a survey of the sky, he came across this small blue ball in the sky and realized that it didn't quite look like a star. So a few days later he observed it again. He had noticed that it had moved in the sky and realized that it wasn't a star at all but it was a planet. That discovery was on March 13 in 1781. At its brightest, Uranus can reach magnitude 5.5 which is barely visible with the naked eye and easily visible in binoculars. Even though it's visible with the naked eye, it had never been recognized as a planet, mostly because it's faint and because of its slow orbit, it moves through the stars very, very slowly. As we searched through historical records we found that Uranus had been observed many times before. At its largest, Uranus is only four arcseconds across. Even through a telescope at high power, it appears as a featureless blue-green ball. Uranus spins on an axis that's tipped 98 degrees with respect to its orbit around the Sun. At times we're looking right down its north pole, at other times we're looking down its south pole, and at other times we're looking over its equator. Because Uranus is so small, and so faint, we can't see any features on the planet.

Similar to Uranus is Neptune. It's a near twin to Uranus. It's 3.8 times the diameter of the Earth. It takes 165 years to go around the Sun and it's 30 astronomical units from the Sun, about six times farther than Jupiter. Neptune, like Uranus, wasn't known to the ancients and it was discovered because of its gravitational influences on Uranus. Its gravity was perturbing, or disrupting, Uranus' orbit and scientists used that to discover where Neptune was in the sky. Like Uranus, you can see Neptune moving through the sky from night to night, but at its best it's only two arcseconds across and it appears as a featureless blue disk in moderate-sized telescopes. Neptune's largest moon, Triton, is magnitude 13.5 and that requires a very large telescope to see. It's not likely that you will see any of the moons of Neptune. But Triton is a good introduction to the last body that we're interested in in the solar system and that's Pluto.

Out at the edge of the solar system there is a belt of cold, icy bodies and Pluto is one of the larger members of this belt of icy bodies called the "Kuiper belt." At magnitude 14.5 Pluto is a challenging object even in a large telescope. It appears as nothing more than a very dim star at the very edge of visibility. And really the only way to see Pluto is to note its motion from night to night. Photograph the stars or do a drawing of what you see through a large telescope and come back a few nights later and one of the stars will have moved, and that will be Pluto.

A number of asteroids are visible through a telescope or a pair of binoculars. Asteroids are called "minor planets" and most of them are small, rocky bodies orbiting between Mars and Jupiter. The largest of the asteroids, Ceres, is 590 miles, or 950 kilometers, across, and it's estimated that there are one to two million asteroids in the asteroid belt. Over 15,000 asteroids have been assigned names and over 200,000 have well-determined orbits. Many of them are bright enough that they're easily visible in a telescope. Even the bright asteroid Vesta, for example, with a mean diameter of about 330 miles, is an asteroid that can reach naked eye visibility at magnitude 5.5. If you observe an asteroid with a telescope you don't see any surface features, you simply see it moving through the sky slowly from night to night.

There's one last group of bodies in our solar system that are very important and these are the comets. But I've left comets out of this talk because bright comets in our sky don't occur very often and I consider them a very special event. In the next lecture we will talk about special events in the night sky, including those rare times that we see an exceptionally bright comet in the evening sky.

# Meteor Showers, Comets, Eclipses, and More
## Lecture 6

An exceptionally bright meteor is called a fireball, and the brightest fireballs can be dazzling … . They can leave a green trail in the sky that lasts from a few seconds to a minute or more. They can even be seen over hundreds of miles. A famous example is the Peekskill meteorite, which burned up over the East Coast of the United States.

A bright comet is one of the most spectacular sights in the night sky, but bright comets generally return to the inner solar system only about once a century. Faint comets, however, are quite common. Comets are made mostly of water ice with some frozen ammonia, carbon dioxide, and carbon monoxide, and they contain chunks of rocks and particles of dust. They are typically about 6 miles across. In their highly elliptical orbits, comets get very close to the Sun and can go as far out as the Kuiper belt or the **Oort cloud**. As a comet approaches the Sun, it warms up and the ice starts to melt, changing to gas. This gas is blown away from the comet by the solar wind, creating two tails: a blue ion tail and a yellow/white dust tail. The tail of a comet can be millions of miles long.

© Digital Vision/Thinkstock.

Halley's comet, a bright comet with a period of 76 years, will next be visible from Earth in 2061.

Comets leave behind tiny particles of rock and dust, and if Earth happens to cross this debris trail, we get a meteor shower. Most **meteors** are about the size of a large grain of sand to a small pebble. They move very fast, about 7 to 45 miles every second. As it streaks through Earth's atmosphere, the meteoroid compresses the gas in front of it, which excites the molecules in our atmosphere to glow.

A solar eclipse at totality.

The bright streak you see in the sky is actually glowing atmosphere as the meteor comes through. The best way to observe a meteor shower is with the naked eye between the hours of midnight and dawn.

A total solar eclipse occurs when the Moon passes between the Earth and the Sun. During **totality**, the outer corona of the Sun can be seen. In the minutes before totality, you will notice that the world around you changes. Shadows of objects take on an odd appearance. The temperature begins to drop, and birds may go to roost. In the final seconds before totality, sunlight can shine through a single valley or crater on the Moon and create a spectacular diamond ring. During totality, you can take off your eye protection and see the solar corona.

Lunar eclipses occur when the Moon passes into the shadow of the Earth. The reason we do not get a lunar eclipse at every full Moon is that the Moon is tilted by about 5° with respect to the ecliptic. During most full Moons, the Moon completely misses the shadow of the Earth. There are three types of lunar eclipses: penumbral, partial, and total. A total lunar eclipse occurs in about three to four hours. Because sunlight bending through the Earth's atmosphere can illuminate the Moon, it usually does not disappear completely during a total lunar eclipse. It takes on a deep red or coppery red color—the combined light of all the world's sunrises and sunsets.

**A total eclipse of the Sun may be the most powerful and emotional astronomical event of your life.**

Other phenomena we can see in the sky include eclipses of Mercury or Venus, bright halos around the Moon and the Sun, satellites, and auroras.

The latter occur when particles are flung off the Sun during a large flare and are caught in the magnetic field of the Earth. When these charged particles collide with the molecules in our atmosphere, they excite the gas to glow. At northern latitudes, the glow is called the aurora borealis, while in the southern latitudes it is called the aurora australis. ■

## Important Terms

**meteor**: A bright streak of light in the sky that appears when a meteoroid enters the Earth's atmosphere.

**Oort cloud**: A giant cloud of cold, icy bodies that stretches tremendous distances away from the Sun.

**totality**: The period during a solar eclipse when the Moon completely covers the Sun.

## Suggested Reading

Brunier and Luminet, *Glorious Eclipses.*

Dickinson, *Nightwatch,* chap. 9–10.

*Heavens-Above.*

Littmann, Wilcox, and Espenak, *Totality.*

Maor, *Venus in Transit.*

*NASA.gov.*

*NASA Eclipse Web Page.*

Ridpath, *Norton's Star Atlas*, chap. 3, p. 88–102.

*The Observer's Handbook.*

Sheehan and Westfall, *The Transits of Venus.*

*Spaceweather.com.*

1.  When is the next total solar eclipse? Go to NASA's eclipse website or look at the maps in the book by Littmann, Willcox, and Espanek and find the date of the next *total* solar eclipse. Follow the links to the map of the path of the Moon's umbral shadow across the Earth. Can you travel to this eclipse? Is there a part of the world that you have wanted to visit that is in the eclipse path? There are many commercial companies and alumni groups that have tours arranged around solar eclipses. Search the Internet for solar eclipse trips.

2.  When is the next lunar eclipse visible? Use NASA's eclipse website to determine when the next total or partial lunar eclipse will be visible in your area. Be sure to mark the date on your calendar.

3.  Are there any bright satellites visible in your sky tonight? Use *Heavens-Above*, *Spaceweather.com*, or a similar site to find out what satellites will be visible in your sky this evening after sunset. Check *Heavens-Above* or NASA's International Space Station pages to see the next time the International Space Station will be visible from your location. Be sure to write the next pass on your calendar and set an alarm so you do not forget.

4.  When is the next good meteor shower that you can see? Use the table in the appendixes to determine when the next good meteor shower will occur in your area. Find a dark sky site and make plans to see the shower. Invite others to join you.

# Meteor Showers, Comets, Eclipses, and More
## Lecture 6—Transcript

There's nothing like being outside at night and being surprised by the bright glow of an exceptionally bright meteor streaking across the sky, or seeing the faint red glow of an unexpected aurora. In this lecture, I'm going to tell you about those special events in the sky that you should be looking for while you're outside at night. Some of these events, like bright meteors and aurora, can't be predicted ahead of time, you just have to be aware that they might occur. But others, like eclipses and annual meteor showers, occur at well known and predictable times, so you can actually make plans to go see one even if those plans require a trip.

A bright comet is one of the most spectacular sights in the night sky. People often ask me, when will the next bright comet be visible that they can see? I'm afraid that most of the time the answer, is I don't know. There are a few dozen bright comets, such as Halley's Comet, which has a period of 76 years, and they return to the inner solar system about once a century or so. That's a long time between passes. Halley's Comet won't be visible again until mid-2061. So you can't always wait for one of these bright comets to reappear.

Faint comets on the other hand are quite common. At any one time there are usually a few visible in small- to moderate-size telescopes. Through the telescope they look like a small, fuzzy patch in the sky. In fact, they look remarkably like a nebula, and the way you can tell the difference between a comet and a nebula is the same way that Charles Messier did. You'll remember that Charles Messier was the 18th century comet hunter who kept finding faint fuzzy objects that didn't move in the sky, and that's the key. Comets, as they orbit around the Sun, are slowly moving, whereas nebula are staying still. As a comet orbits the Sun, it moves from night to night, and the motion is clearly visible if you observe a comet tonight and then tomorrow night you will definitely see that it has moved. And with some of the closer comets that are moving faster through our sky, you can actually see the motion in just a few hours.

Every few years a comet gets bright enough that you can see it with the naked eye. Most of these appear as a faint fuzzy patch in the sky but they probably won't be visible in light-polluted skies unless you use a pair of binoculars or a telescope. The real treat is those very bright comets and those are rare and they occur only a few times in the course of a human lifetime. Usually when I talk to people about a bright comet in the sky, they'll relate to me to that time, a decade or so earlier, when they saw a bright comet and its beautiful tail in the sky. Bright comets have even affected history. In ancient and medieval times, a bright comet was seen as a bad, if not outright terrible omen, for the ruler or king. For example, in 1066 A.D., Halley's Comet appeared in the sky as William of Normandy defeated the Saxon king, Harold II. In fact, the very first image that we have of Halley's Comet appears on the Bayeaux Tapestry that tells the story of that defeat.

Comets are big, giant, dirty snowballs. They're made mostly of water ice with a little bit of frozen ammonia, frozen carbon dioxide, frozen carbon monoxide in them as well. And they also contain lots of chunks of rocks and small particles of dust. They're typically six miles, or 10 kilometers, across. Halley's Comet, for example, was last back in the inner solar system in 1985 and 1986. This is a picture of Halley's Comet taken by the European Space Agency's Giotto spacecraft as it flew by the nucleus or the cold dirty snowball that makes up the center of Halley's Comet. Comets orbit the Sun on highly elliptical orbits. They get very, very close to the Sun, often coming closer than Venus and sometimes closer than Mercury to the Sun, and then at their farthest, they go far out in the solar system. Most of these very bright comets have periods that are measured in thousands of years. They haven't been seen before in recorded human history and it'll be thousands of years before they return to the inner solar system as well.

Some of these comets go as far out as the Kuiper Belt. You'll remember from the last lecture that's that cold belt of icy bodies that include Pluto and Aris. Some of them go as far out as the Oort Cloud, which is a giant cloud of cold icy bodies that stretches tremendous distances away from the Sun. As a comet approaches the Sun it warms up, and as the comet warms up the ice starts to melt. But unlike on Earth, the ice doesn't turn into a liquid. Because it's in the vacuum of space, the ice goes directly into a gas. This gas gets blown away from the comet by the solar wind, and it creates two tails. Typically a blue ion tail, which glows for exactly the same reason that a fluorescent bulb does. In the case of the ion tail, the atoms that make up the tail are excited to glow by the ultraviolet light from the Sun. The dust tail, on the other hand, is typically yellow-white in color, and that's because it comes from little tiny particles of rock and dust that have been shed off the comet. And these particles of rock and dust are simply reflecting sunlight and so they appear the same color as the Sun, or yellow and white.

The tail of a comet can be millions of miles long, and the tail always points away from the Sun because it's this gentle wind from the Sun called the "solar wind" that blows that tail outward. The solar wind is just made of up tiny little particles of gas that are gently blowing off the Sun and they push the little tiny particles of rock and dust and the gas that comes off the comet straight away from the Sun. What you'll see through a telescope on a comet is a round, fuzzy patch at the center of the comet called the "coma." The coma is actually much bigger than the nucleus. The nucleus is the icy body itself, but since that icy body is so small and it's very, very faint, it normally isn't visible in a telescope. You might see at the center of the coma a very bright center and the nucleus is buried deep down inside of there. On most bright comets you should be able to see the dust tail stretching away from the coma. The ion tail, the blue tail, though, can be very difficult to see because it's very faint. Through a telescope you may also see jets of materials. In that image of Halley's Comet you may have noticed that the material doesn't shed gently but there are places where the material is literally jetting off the surface and sometimes you can see these jets in a telescope. Occasionally, in the tail itself, you will see a knot or a bright streamer that moves as the solar wind blows it away from the comet.

Comets leave behind them little tiny particles of rock and dust, and if the Earth happens to cross this debris trail that the comets have left behind, we get a meteor shower. On a clear night, from a dark location, you can expect to see a few to maybe up to 10 meteors per hour. They've been called "shooting stars" but it's important to remember that they have nothing at all to do with stars. They are not falling from the sky. They're actually made up of little tiny particles of rock and ice. Before they burn up in the sky they're called "meteoroids," and as they burn up they become "meteors." They're surprisingly small for that bright streak you see in the sky. Most meteors are about the size of a large grain of sand to a small pebble. They're moving very, very fast, about seven to 45 miles every second, that's 11 to 72 kilometers per second, and that means they have lots of energy. As they streak through the Earth's atmosphere, friction with the atmosphere heats them up and it heats up the Earth's atmosphere. As they plow through the atmosphere they compress the gas in front of them, and compressing a gas makes it hot, and this excites the molecules in our atmosphere to glow. So that bright streak you see in the sky is actually glowing atmosphere as this meteor comes through. Most of these little tiny particles burn up about 50 to 60 miles above the surface of the Earth, which is about 80 to 100 kilometers. Most of the meteors that you see at night are sporadics, which means just random particles of rock and dust hitting the Earth.

An exceptionally bright meteor is called a "fireball," and the brightest fireballs can be dazzling. They can cast a shadow on the ground. They can leave a green trail in the sky that lasts from a few seconds to a minute or more. They can even been seen over hundreds of miles. A famous example is the Peekskill Meteorite, which burned up over the East Coast of the United States. In this picture from a movie you can see this meteorite as it breaks up into multiple pieces and some are even seem to flash brightly or explode. If the meteorite is big enough, some pieces of it will make it all the way down to the surface of the Earth. In fact, in this meteor, the Peekskill Meteorite, a large piece actually landed on the rear of a car in New York and did significant damage to the car. Meteorites have been known to punch holes in the roofs of buildings, and even a few people have been hit after the meteorite passed through the roof slowing them down. So far we don't know of anyone who has ever been killed by a falling meteorite.

On the other hand, if the Earth passes through a dense trail left by a comet we actually get a meteor shower. There are about 10 good meteor showers every year. The meteors appear to fan out or radiate from a single spot in the sky called the "radiant." All the meteors are traveling on parallel paths with one another, and just like a pair of railroad tracks appears to converge off in the distance, all these meteors traveling on parallel paths appear to converge in the sky. A meteor shower gets its name from the constellation where the radiant is located, where all the meteors seem to be streaking out of. The Perseids, which are visible in August, appear to radiate from the constellation Perseus. The Leonids of November come from Leo. A measure of how many meteors you will see per hour in a shower is the zenith hourly rate, and as the name implies it's the number of meteors you would see per hour high up at the zenith where you have the least amount of interference from the atmosphere.

Here's a list of the good annual meteor showers and just a couple of them are worth pointing out. The Perseids, which occur from July to August of every year, typically see about 80 meteors per hour. Most meteor showers, although they stretch over a couple of weeks, are best on one night. I've listed in this table the maximum night, the best night for going out and seeing this shower. In the case of the Perseids, and most meteor showers, this can vary by one day before or one day after that listed in the table depending on how the Earth's orbit intersects the shower of the meteors or the former orbit of the comet. You can look up in magazines and on the Internet for a given year exactly what day is the best night to go out. In the case of the Perseids, as I said, you'll see about 80 meteors per hour and I've also listed in the table where in the sky to look for it. They're radiating from the constellation Perseus, which its right ascension and declination is given there, and on the very right-hand side of the table I've listed the parent body. In the case of the Perseids it's coming from Comet Swift-Tuttle. Another excellent shower is the Geminids in early December, typically peaking around December 13th, and there you see about 100 meteors per hour. And that comes not from a comet actually but from an asteroid, the minor planet Phaethon.

The best way to observe a meteor shower is actually with the naked eye. You need to go out and find dark skies because a bright Moon can interfere. The brighter the Moon the fewer meteors you will see. The fields of view of binocular or telescopes are just too small to see any meteors, so don't try using them. You're looking at such a small part of the sky that the odds of a meteor passing through the telescope or binoculars are very, very small. Use a comfortable chair, lie down on the ground in a sleeping bag or a blanket, and just look up in the sky with your naked eyes. It's best to face towards the radiant and look about halfway up in the sky to see the most meteors. For the best showers you'll expect to see about one meteor per minute.

Most meteor showers are best from midnight until dawn because that's when you're on the side of the Earth facing the direction of our travel around the Earth. You'll remember this image from earlier in the course. You see people located at four different times of day. The motion of the Earth here is carrying us downward as we orbit around the Sun. That is, when you're standing at sunrise you are facing the part of the Earth that's towards the Earth's motion. We see the most meteorites, or most meteors in the sky, from midnight to noon because then we're on the side of the Sun facing towards the direction in which Earth is moving. I think the best analogy to this is the bugs on the windshield. When you're driving on the highway all the bugs end up on the front windshield because you're literally collecting them as you drive along. The same is true with meteors. They're most visible from midnight to noon because those are the ones that we're running into.

Comets are not the only important astronomical object that influenced human history. Equally important were total eclipses of the Sun. In fact, many rulers over the ages had court astronomers specifically to tell them when solar eclipses would occur. This was particularly important if the ruler happened to have his power derive from the Sun. A total eclipse of the Sun may be the most powerful and emotional astronomical event of your life. For me personally, my two most memorable astronomical experiences were the two total solar eclipses that I've been fortunate enough to see. It's a wonderful coincidence in our sky that the Moon and the Sun are each half a degree across, which means that every now and then the Moon can cover the Sun and completely cover it up.

A total solar eclipse occurs when the Moon passes between the Earth and the Sun. The shadow of a body consists of two parts. The umbra is the darkest part of the shadow. It's where the Sun is completely covered by the Moon, and surrounding that is the penumbra where the Sun is not completely covered but only partially covered. In a solar eclipse, the small spot of the Moon's umbra will fall on the Earth, and if you're lucky enough to be in that small dark spot, you will see the Moon completely cover the Sun. Over the course of an hour, the Moon will gradually cover the Sun bit by bit until the Sun is completely blocked from view. During totality, which is that time when the Moon completely covers the Sun, you can see the outer corona of the Sun. It's this faint, hot, outer atmosphere of the Sun that's normally completely invisible from view. If you're not lucky enough to be inside the umbra, and you happen to be in the penumbra outside the umbra, you'll see a partial solar eclipse where the Moon covers only part of the Sun.

Safety is paramount when you're observing a total eclipse of the Sun. It's important to say that the Sun is no more dangerous during an eclipse than it is at any other time. To view the partial phases, though, leading up to totality, you'll need eye protection. In Lecture 4, we talked about ways to safely observe the Sun. During the partial phases of the eclipse you could use a pair of eclipse glasses that will let you safely look at the Sun. Or you could use one of these projection telescopes like the Sunspotter or the solar scope to see the Moon gradually coming in front of the Sun. In the minutes before totality, as you look at the world around you, you'll notice the world changing. Shadows of objects will take on an odd appearance. If you happen to have a tree nearby, little pinholes of light shining through the tree, or artificially-created pinholes by poking a hole in a piece of paper, you will see that the sunlight shining through these little pinholes looks like little tiny crescents.

Just before totality the sky starts to turn dark, and this is the shadow of the Moon approaching. The temperature begins to drop. You may notice odd things around you. Birds may go to roost. Animals can change their behavior. In the final seconds just before totality, sunlight can shine through a single valley or crater on the Moon and create a spectacular diamond ring. The diamond ring is caused because the bright sunlight shining through that one little valley makes the bright spot, while the glow, the ring part, comes from the Sun's corona appearing around the Moon. Suddenly as the Moon covers the Sun, you're plunged into darkness and the Sun disappears. Now, and only now, it's safe to take off the glasses and look at the Sun with your eyes. Planets and bright stars become visible, and all around the horizon there's light shining on the horizon just as if it was sunset or sunrise all around you because in fact the Sun is shining out there hundreds of miles away.

During totality, as I said, you can take off your glasses and see that solar corona around the Moon. The corona has a temperature of about a million degrees Celsius, and it's never visible to the eye at any other time than during a total solar eclipse. The corona is one of the most beautiful things that you will see during a solar eclipse. It has streamers and loops of material and it's really quite remarkable. If you're lucky, you may even see a prominence, which is a loop of gas flung off of the surface of the Sun poking up just above the limb of the Moon.

Not all solar eclipses are total. The Moon has to be big enough to completely cover the Sun. And the Moon's orbit around the Earth is elliptical, which means sometimes it's closer to us and appears larger, and sometimes it's farther away and appears smaller. The same is true, by the way, of the Earth's orbit around the Sun. As we saw earlier, there are times of the year when the Earth is closest to the Sun and times when we're farther away. When we're close to the Sun it looks bigger and when we're farther away it looks smaller. There are times when a solar eclipse occurs when the Moon happens to be just too small to cover the Sun. This typically happens when we're closest to the Sun and the Sun is big and the Moon is far away from us and it appears too small. If the Moon isn't big enough to completely cover the Sun we get an annular eclipse, where a small ring of light from the Sun's atmosphere remains visible around the Moon. Even though 99 percent or more of the Sun is blocked from your view you can't see the corona, you can't look at it with your naked eye, and you have to use all of these safety precautions for observing the Sun. None of the spectacular events that happen during a total solar eclipse are visible during an annular eclipse.

On average, the Moon is too small to cover the Sun, so annular eclipses are actually more common than solar eclipses. A total solar eclipse occurs about once every 18 months somewhere in the world. That shadow spot is very small, so it's most likely that you're going to have to travel to see the eclipse. From any one given spot, a total eclipse occurs about once every 360 years. At the equator, the size of the Moon's umbra, is at best, 169 miles across, which is 273 kilometers. Because of the orbital motion of the Moon around the Earth, and the rotation of the Earth, this tiny spot is moving across the Earth at over 1,000 miles per hour, 1,700 kilometers per hour. The total length of time that the Moon covers the Sun is no more than seven minutes and 31 seconds. If you would like to see a total solar eclipse, you'll have to make plans well in advance. There are many ways that you can fly. A number of cruise companies offer cruises to go see total solar eclipses. You'll have to find a map of the Earth that shows you the path of eclipses across the surface and plan out when and where you might get to see one of these most spectacular events in our sky. It sounds crazy to travel to one of these eclipses, but everybody I know who has seen a total solar eclipse is already planning their next trip to go see one.

Most people have not been fortunate enough to see a total solar eclipse. There is an eclipse though that everyone has the opportunity to see. Though it's much less common than solar eclipses, they can be seen by more people. Lunar eclipses occur when the Moon passes into the shadow of the Earth, and you might have wondered why it is that we don't get a lunar eclipse at every full moon. That's because the orbit of the Moon is tilted by about five degrees with respect to the ecliptic.

During most full Moons, the Moon actually passes below the umbra of the Earth or passes above the umbra of the Earth, in fact, passes below the penumbra and above the penumbra of the Earth, so that the Moon completely misses the shadow of the Earth. In fact, it's only twice a year, during eclipse seasons, when the orbit of the Moon lines up with the ecliptic, the path of the Sun in the sky, when an eclipse can occur. In fact, this is the origin of the name "ecliptic." You might have wonder why it is that we call the path of the Sun through the sky the "ecliptic." The answer is, only when the Moon appears close to the ecliptic at full can a lunar eclipse occur.

There are three types of lunar eclipses: penumbral, partial, and total. A penumbra lunar eclipse is when the Moon passes through the penumbra of the Earth. Remember the penumbra is the part where only part of the Sun is blocked from view. This is not the deepest, darkest part of the Earth's shadow. During a penumbral lunar eclipse it's actually difficult to see any difference in the Moon at all. Typically penumbral eclipses are not worth watching, because the Moon doesn't pass into the dark umbra of the Earth. A partial lunar eclipse occurs when only part of the Moon passes into the umbra, so the Moon will not completely go inside the umbra. There will always be a small sliver of the Moon that's still brightly lit up by the sunlight, but you get to see the Moon as it passes through the shadow.

A total lunar eclipse happens when the Moon completely passes into the umbra of the Earth. Over the course of about an hour or so, the Moon disappears into the shadow of the Earth. It can be completely inside the shadow of the Earth, the umbra, for about one hour and 47 minutes, and then it takes about another hour for it to come out. In the space of three to four hours, the Moon goes from being full, to being fully eclipsed, to being full again.

Sunlight bending through the Earth's atmosphere can illuminate the Moon, and so during a total solar eclipse the Moon doesn't usually disappear. It takes on a deep red or a coppery red color. This is because that light that's reaching the Moon is the combined light of all of the world's sunrises and sunsets. That light has to travel through so much of the Earth's atmosphere that the blue light is filtered out and only the red light makes it to the Moon. How dark the Moon gets during an eclipse depends on how much dust there is in the atmosphere and how much cloud cover there is around the limb of the Earth as seen from the Moon. Astronomers judge the darkness of an eclipse on the Danjon scale. An L0 Danjon scale eclipse is a very dark eclipse where the Moon is almost invisible, and it scales all the way down to an L4 where the Moon is still very bright but coppery red or orange in color. A lunar eclipse is visible to everyone on the night side of the Earth. There are two to four lunar eclipses of some type every year. On average, a lunar eclipse is visible from any location on Earth about once every 18 months. So it's not at all difficult to see a lunar eclipse.

It's a very common misconception that Christopher Columbus was the first person to prove that the Earth was round. Actually nearly 2000 years earlier Aristotle used the apparent shape of the Earth's shadow to show that the Earth is a sphere. Aristotle noticed that during total solar eclipses, as the Moon passed into the shadow of the Earth, the shadow of the Earth was always round. Aristotle realized that the only body that always casts a round shadow is a sphere. He also had other lines of evidence. Aristotle knew, as we've seen in this class, that as you move north or south of the equator the stars in the sky shift back and forth. That will only happen as you walk on a curved surface. He also noticed that as a ship disappears over the horizon it disappears from the bottom up as it sails away. If the Earth were flat, it would just get smaller and smaller. Aristotle had well established by around 350 B.C. that the Earth was spherical or round.

This is a list of upcoming eclipses, total and partial lunar eclipses. The dates are given on the left side, the type of the eclipse in the middle, and the locations where they're visible is given over on the right side. You can use this table in planning when you might next be able to see a lunar eclipse in your sky.

An eclipse of another type occurs when either Mercury or Venus passes in front of the Sun. These eclipses are much more rare. Since both Mercury and Venus are closer to the Sun than the Earth, they can, on occasion, pass directly in front of the Sun. You might think this happens on every orbit. But just like the Moon, the orbits of Mercury and Venus are tilted by about seven degrees for Mercury and 3.4 degrees for Venus. So usually they pass above or below the Sun. Every now and then, when they're between the Earth and the Sun, they can pass directly in front of the Sun. During one of these transits we will actually see a small black spot on the Sun, which is the shadow of the planet. Mercury is small enough, the transits of Mercury require a telescope to see.

Here's a picture showing Mercury against the Sun. You'll see a dark sunspot over on the right-hand side much larger than the small planet Mercury. The orbit of Mercury allows the planet to transit the Sun 13 or 14 times every century. These transits are fairly rare and you have to plan ahead of time when one is occurring so that you'll be able to have a solar telescope ready to go out and see it. During this transit Mercury is very small, only 10 to 12 arcseconds across.

Transits of Venus are even more rare. They occur in pairs and the transits in a pair are separated by eight years, but the pairs themselves are separated by 105.5 or 121.5 years. Transits of Venus only occur in December or early June when the Earth crosses Venus' tilted orbit. In the 21st century the transits are lasting about six hours. Here's a list of the transits of Venus so that you can see how exceptionally rare this event is. During a transit of Venus, Venus is large enough that with solar glasses you can see the spot of Venus crossing the Sun. Then with a telescope, it will show you the spot in even more detail.

Another question I get all the time, especially on a full Moon night during the winter, is, what is that bright halo around the Moon? Did something happen to the Moon? Bright halos around the Moon and the Sun are quite common and they occur when high, thin cirrus clouds filled with ice crystals are covering our sky. These clouds are typically three to six miles, or five to 10 kilometers, in height, and they have lots of long hexagonal ice crystals inside of them. As the light from the Moon or the Sun comes through these ice crystals, the light gets bent by 22 degrees or more. The crystals that are more than 22 degrees from the Sun or the Moon will scatter some of the light into your eyes creating this bright halo in the sky. We not only see Moon halos in the sky but occasionally with the Sun we can see sun phenomena caused by these ice crystals as well. Probably the most common one are sun dogs. Sun dogs are two bright spots 22 degrees to the left and 22 degrees to the right of the Sun. They usually have a red tinge on the side that's closest to the Sun and they can appear anywhere at any time.

Under exceptional conditions we might also see a bright, rainbow-colored arc called the "circumzenithal arc" high overhead and then rarely you would see things such as the tangent arc or the parhelic circle. At times you can see the full 22 degree halo around the Sun. It's brightest along the inner edge which has a reddish color. The same ice crystals in these cirrus clouds that are responsible for the 22 degree halo around the Sun are the ones that are responsible for that 22 degree halo around the Moon. As I said earlier, this is most obvious when the Moon is full because then it's giving off the most light.

It's quite common just after sunset to see a half dozen or more satellites crossing the sky in the few hours after sunset and before sunrise. Most satellites look a dim- to moderately-bright star moving slowly across the sky. But some of them, like the International Space Station, can be brighter than Venus. The satellites are simply shining by reflected sunlight.

Although the Sun has set for you and it's dark where you are, the satellites themselves are in full daylight. Satellite passes are visible for a few hours after sunset and a few hours before sunrise. They're never visible in the middle of the night because just as we're deep inside the Earth's shadow in the middle of the night, the satellites are as well. A typical satellite pass takes about five to ten minutes and it's important to know that most of the objects you're seeing are not functioning satellites. Most of them are the upper stages of rockets that were used to launch the satellites and those are called "rocket bodies."

You can predict which satellites will be visible in your sky using any one of a number of webpages on the Internet, including NASA.gov. You can print not only when they'll be going over head but what path they will take through the sky. It's not worth trying to spot a satellite unless it's going to climb more than 20 or 30 degrees above the horizon because below that altitude they're dim and difficult to see and any trees or buildings will block your view.

The vast majority of satellites that you can see with the naked eye are in low Earth orbit, typically 200 to 600 miles above the surface, and in order to stay in orbit they're moving about 17,500 miles per hour. If you're lucky enough to see the International Space Station, which typically has a good pass from most locations a couple of times a month, just keep in mind as you see that bright light moving across the sky that there are people inside, living and working there.

The last thing that I want you to look for in the sky are aurora. Every now and then, particles from the Sun during a large flare will be flung off the Sun and they'll get caught in the magnetic field of the Earth. The magnetic field from the Earth will funnel these particles into the North and South Poles. When these charged particles collide with the molecules in our atmosphere they excite the gas to glow, just like a fluorescent bulb or like the meteors that we talked about earlier.

At northern latitudes the glow is called the "aurora borealis" while in the southern latitudes it's called the "aurora Australis." Although it's common to see the aurora at high northern or southern latitudes, it's rare to see it at low latitudes or even mid-latitudes. Generally you will only see an aurora if a major solar flare occurs. At high latitudes where the aurora are common, they take on a variety of shapes and colors. They can look green or red and they can even appear as giant curtains in the sky. But at mid latitudes they look much more like a diffuse red glow in the sky. Typically at mid latitudes we'll only see them during a giant flare on the Sun.

In the next lecture we'll be talking about the Northern Sky, those constellations that might be visible behind that bright aurora that you see in the Northern Sky. It's a lecture that we might come back to again and again to see over and over because these constellations are visible year round.

# The Northern Sky and the North Celestial Pole
## Lecture 7

**M82 is an example of a starburst galaxy undergoing a tremendous burst of star formation. In our Milky Way, stars are forming at a rate of about one sun-like star every year. In M82, the star formation rate is about 10 times higher.**

Regardless of when you go outside, Ursa Major, the Big Bear, and Cassiopeia are always visible in the sky. We'll use them in this and ensuing lectures to find other constellations.

We begin with the most familiar shape in the northern sky, the Big Dipper, the position of which changes with the seasons. Any time of the year, we can follow the two end stars of the Big Dipper, called the pointer stars, to find our way to the North Star, Polaris. In the fall, we can draw a line from the handle of the Big Dipper through Polaris, across the **North Celestial Pole**, and come to the distinct *W* shape of Cassiopeia. If we follow the two bright stars on the western side of Cassiopeia, we come to her husband, Cepheus. In the summer, if we draw a line from the bowl of the Little Dipper to the bright star Vega, we find the head of Draco the Dragon.

**Both the Greeks and the Native Americans, two civilizations widely separated by half the Earth, saw a bear in this part of the sky.**

In the first lecture, we saw that the sky around us is in constant motion owing to the rotation of the Earth. Not only do the Sun and the Moon circle the sky each day, but the distant stars rise in the east and set in the west. The North Celestial Pole is the axis around which the sky appears to rotate from east to west. Polaris is close to, but not exactly at, the North Pole on Earth. Some stars circling the North Celestial Pole are always above the horizon and never rise or set. These are called **circumpolar** because they circle the sky all day and all night.

As you travel around the Northern Hemisphere, try to find Polaris wherever you go. At high latitudes, it will always be high overhead, but at low latitudes, it will always be close to the horizon. The altitude of the pole and your latitude are always the same, a fact that is significant for navigation. As the Earth wobbles, the place in the sky where the pole is pointing shifts around. A single wobble of the Earth takes about 26,000 years. Thus, Polaris has not always been, nor will it always be, the pole star. Another consequence of this shifting (precession) is that the equinoxes shift with respect to the background stars. Today, the vernal equinox is located in the constellation of Pisces.

Ursa Major, the Big Bear, is the brightest and easiest constellation to find in the northern sky, while most of the stars of Ursa Minor, the Little Bear, are faint. Both the Greeks and Native Americans have wonderful stories to explain why these two bears have long tails in the sky. In and around Ursa Major are some of the best nearby galaxies in our sky. For example, just southwest of Alkaid, the end star in the handle of the Big Dipper, lies M51, a grand spiral galaxy about 31 million light years away.

In terms of area, Draco the Dragon is one of the largest constellations in the sky. Its head is about halfway along a line from the bowl of the Little Dipper to the bright star Vega. One of the most famous stars in the sky is delta Cephei in Cepheus; it's an unstable, pulsating, variable star about 890 light years away. Such Cepheid stars can be used to measure distances to other galaxies. ∎

## Important Terms

**circumpolar**: Refers to stars that never set.

**North Celestial Pole**: The point directly above the Earth's North Pole projected into space.

## Suggested Reading

Dickinson, *Nightwatch*, charts 1–2.

Millar, *The Amateur Astronomer's Introduction to the Celestial Sphere*.

Monroe and Willamson, *They Dance in the Sky*.

Ridpath, *Norton's Star Atlas*, charts 1–2.

Schaaf, *A Year of the Stars*, 138–148, 186–187, 279–282.

## Questions to Consider

1. If Ursa Major and Cassiopeia are circumpolar, are they visible every day of the year? Use your planisphere or star maps to find when Ursa Major and Cassiopeia will be visible in the sky during the year.

2. Can you find other stories about Ursa Major? Ursa Major may be the most important constellation in the sky. Not only does it have a distinctive shape with moderately bright stars, but it is also visible year round and is useful for finding north. Thus, nearly every civilization in the Northern Hemisphere has created a constellation or constellations out of some of the 7 stars of the Big Dipper. Learn something about the mythological figures and constellations that other cultures have placed in this part of the sky.

3. Is there a time when we will have a better pole star than today?

4. How does the shifting of the vernal equinox in the sky affect the location of the Sun with respect to the constellations? Your birth sign, or astrological sun sign, is supposed to represent the constellation that the Sun was in when you were born. Use an astronomy magazine or research the dates when the Sun is in each constellation. Was the Sun in your birth sign constellation on your birthday? If not, why not?

# The Northern Sky and the North Celestial Pole
## Lecture 7—Transcript

In this lecture we're going to talk about the stars and the constellations that are visible around the North Pole and our sky. Regardless of when you go outside, Ursa Major, the Big Bear, and Cassiopeia, the queen of the mythical kingdom of Ethiopia, are always visible in the sky. Not only will we use them in this lecture to find other constellations, we'll come back to these two in future lectures as a starting point for navigating the seasonal skies.

Thus, this is a lecture that you might want to return to time and again to review these familiar constellations. In each of the next six lectures, where we tour the sky, I'll begin with a short tour of the sky so that you can see where the constellations are relative to one another and their positions in the sky. You'll want to have a star map or planisphere handy so that you can follow our route.

After the tour, we'll consider each constellation in turn, and talk about the stars and other objects that are interesting in each of the constellations. We'll begin today with the most familiar shape in the Northern Sky, the asterism of the Big Dipper, or Plough. You might remember from Lecture 1 that the Big Dipper is made up of seven stars. Three stars form the handle of the dipper and four stars form the bowl of the dipper. It looks like a spoon or a dipper in our sky.

In Great Britain they call it the "Plow" and in other parts of Europe it's known as the "Wagon" with the four stars of the bowl forming the wheels of the wagon and the three stars of the handle forming the horses pulling the wagon through the sky. Since the Big Dipper is made of bright stars and it's easy to find, we'll use it to locate the other constellations in the Northern Sky. For most of these lectures, we'll assume that you're out watching the stars at about 10:00 o'clock in the evening.

The position of the Big Dipper changes with the seasons. If you have a planisphere set it for April 15 at 10:00 p.m. In the northern spring, the Big Dipper is high in the sky in the evening. It's the same view that you'll get on March 15 at midnight, or May 15 at 8:00 in the evening. By summer, say July 15th, the Big Dipper is descending down in the northwest.

Now you might remember from your planisphere that east and west are reversed because planispheres are meant to be held up to the sky. For that reason northwest will be on the upper right-hand side of the planisphere, so once you know that you'll see the Big Dipper there on the upper right-hand side. Keep in mind that the best way to use a planisphere is to hold the direction that you're facing at the bottom of the planisphere. If you face north and hold north down at the bottom, you'll see that Ursa Major is relative to its position in the sky.

By the fall, say October 15th at 10:00 p.m., the Big Dipper is low along the northern horizon in the evening. Generally just after sunset in the early evening in the fall we don't get to see the Big Dipper because if you have any trees along your northern horizon, it might be hidden from view. It won't completely set for most people at mid-northern latitudes. In the wintertime, the Big Dipper is rising in the northeast in the early evening and so it's just starting to come out of the trees and later on in the night will be high overhead.

Any time of the year we can follow the two end stars of the Big Dipper, called the "pointer stars," to find our way to the North Star, Polaris, which is about 28 degrees away. Remember that your arm held out at arm's length is 20 degrees, so 28 degrees is your hand held out and then another fist which is 10 degrees. Polaris is very close to the North Celestial Pole in the sky. Many people are surprised when they first see Polaris because it's only a second magnitude star. Many people are under the impression that it actually should be a much brighter star because it's so important in our sky, but it really is a rather dim star.

Polaris is the end star in the Little Dipper. Three stars form the handle of the Little Dipper and four stars form the bowl of the Little Dipper. Like the Big Dipper, the Little Dipper is an asterism. The constellation itself is Ursa Minor, the Lesser Bear.

In the fall, we can draw a line from the handle of the Big Dipper, through Polaris, across the pole and come to the distinct "W" shape of Cassiopeia, the queen of the mythical kingdom of Ethiopia. Depending on the time of night and the time of year, the "W" could be oriented as an "M," or as a "3" or even as an "E." If we follow the two bright stars on the right side, which is the western side of Cassiopeia, we come to her husband, Cepheus, the king of the mythical kingdom of Ethiopia. Cepheus looks very much like a stick figure drawing of a house.

In the summer, if we draw a line from the bowl of the Little Dipper to the bright star Vega, about halfway in between, we find the head of Draco the Dragon. Draco has a long, thin body that curves from the head and the space between the Big and Little Dippers.

In the first lecture, we saw the sky around us is in constant motion due to the rotation of the Earth. Not only do the Sun and the Moon circle the sky each day, but the distant stars rise in the east and set in the west. The Earth is turning on an axis that goes through the North and South Poles. If we project the North and South Poles of the Earth into space onto the celestial sphere, we find a point called the "North Celestial Pole."

The North Celestial Pole is the axis around which the sky appears to rotate from east to west. In this time exposure photograph, we see the stars circling around the North Celestial Pole in the sky. The stars, as they rise in the east and set in the west, are moving counterclockwise as seen in this picture.

When you use the Big Dipper to find Polaris, you get close to, but not exactly at, the North Pole. Polaris is about 0.7 degrees away from the pole. Polaris is not exactly at the pole, you can see in this image how it, too, makes a little tiny circle around the North Pole in the sky. But for most purposes, Polaris is close enough to the pole that it will do just fine. You can find north by locating Polaris, and dropping a line straight down to the horizon. Where the line meets the horizon is due north.

Since ancient times, Ursa Major and Ursa Minor have been used to find north for navigation. For example, Homer, around 850 B.C., tells us that sailors were already steering ships using the stars, that Odysseus kept the celestial bear on his left in order to sail to the east. More recently, in the 19th century in the United States, slaves pictured Ursa Major as a drinking gourd. Escaping slaves were told to "follow the drinking gourd," which would lead them to north and to freedom.

There is, of course, a celestial pole in the Southern Hemisphere as well. But, unfortunately, around the celestial pole in the Southern Hemisphere the stars circle round and round and they never rise for those of us here in the Northern Hemisphere. We'll see in Lecture 12 how to use bright stars in the Southern Sky to find the celestial pole since there's no bright star that marks the South Pole in the sky. At mid-northern latitudes, that South Celestial Pole is always below the horizon.

A good example of one of the sets of stars that circle around the South Celestial Pole that never rise up for those of us at mid-northern latitudes, for example, most people in Europe, the United States, and Canada, is the Southern Cross. The Southern Cross is one of the most beautiful sights in the sky and it never rises above the horizon for people at mid-northern latitudes. It always circles round and round the South Celestial Pole never coming up above the horizon. You don't see any of the Cross above a latitude of 33 degrees north and you have to be below a latitude of 27 degrees to see the whole Cross, but even then, it just skims the southern horizon.

Just as there are stars that are circling the South Celestial Pole that we never see, there are stars circling the North Celestial Pole that are always above the horizon and never rise or set. Since these constellations circle the sky all day and all night, they're called "circumpolar." When you find Polaris at night, keep in mind that Polaris is up there in the exact same position during the day, just as it is at night, you just can't see it because it's invisible with the bright day sky.

For example, take your planisphere and set it for 10:00 p.m. on April 15. You will see the Big Dipper and Ursa Major high in the sky near the center of the planisphere, the center of the planisphere is the point high over your head. If you rotate the planisphere to 11:00 p.m., and then midnight, and then 1:00 a.m., and then 2:00 a.m., and keep going you'll see Ursa Major circling around the North Celestial Pole. If you watch all night long you will see that the stars in the bowl of the Dipper never set.

As you travel around the Northern Hemisphere, make an effort to find Polaris wherever you go. At high latitudes it will always be high overhead, but at low latitudes it will always be close to the horizon. That means that there's a very important relationship between the altitude of the pole and your latitude.

Imagine standing on the North Pole of the Earth at a latitude of 90 degrees. The North Celestial Pole will be directly overhead, at an altitude of 90 degrees. Note that at this location on the North Pole the altitude of the pole, how high up Polaris appears, it's 90 degrees up, and your latitude, 90 degrees north, are the same. Imagine that I walk on the curved surface of the Earth about 10 degrees away.

As I move south along the curved surface of the Earth, Polaris will no longer be overhead. It will move about 10 degrees away if I move 10 degrees to the south. Now that I've shifted myself 10 degrees south, I'm at a latitude of 80 degrees, and if Polaris shifts from being overhead to 10 degrees away from being overhead, it's at an altitude of 80 degrees.

Again my latitude and the altitude of the pole in the sky are the same. By continuing this exercise we see that the altitude of the North Celestial Pole in the sky is identical to your latitude. This is incredibly important for navigation. In fact this is how navigators used to measure their latitude on the surface of the Earth. If you go outside and measure the altitude of the North Celestial Pole in the sky, the altitude of the North Celestial Pole is equivalent to your latitude.

As we heard in Lecture 2, the great Greek astronomer Hipparchus compiled a catalog of the bright stars in the sky around 150 B.C. And when Hipparchus compared his star positions to those of a previous astronomer, he discovered that the stars in the sky had shifted their positions in the sky. Today we realize that this shifting is due to the shifting of the poles in the sky, called "precession." Not only is the Earth spinning around, it's wobbling like a top and we call this wobble "precession."

As the Earth wobbles, the place in the sky where the pole is pointing shifts around. A single wobble of the Earth takes about 26,000 years and as the Earth wobbles, both the North and South Celestial Poles in the sky are shifting around. Thus, Polaris has not always been, nor will it always be, the Pole Star. This diagram shows the shifting position of the North Celestial Pole in the sky. Around 4,500 years ago, around 2500 B.C., when the Egyptians built the pyramids, the dim star Thuban, in Draco the Dragon, was the pole star.

About 12,000 years from now in 14,000 A.D., the bright star Vega will be near the pole. We happen to live in a very special age where we have a bright pole star, Polaris. Precession right now is carrying Polaris slightly closer to the pole. At this time it's about 0.7 degrees from the pole and around the year 2100 it'll be closest to the North Celestial Pole, just under half a degree away.

Note in that diagram that the pole circles a point in the bend of Draco the Dragon. As the Earth orbits the Sun in a flat plane, if we draw a line perpendicular to that plane, it intersects the sky at the north and south ecliptic poles. Remember the Earth's orbit around the Sun is called the "ecliptic." If we take the pole of that orbit it's called the "north" and "south ecliptic poles." Essentially they're just the poles of the Earth's orbit.

As we saw in Lecture 4 on the movement of the Sun in the sky, the Earth doesn't orbit with its axis straight up and down. It's tilted over by about 23.5 degrees. That is as the Earth goes around the Sun the North Pole isn't pointing at the ecliptic pole, it's pointing 23.5 degrees away. As the Earth wobbles every 26,000 years, the North Pole circles around the north pole of the Earth's orbit in the sky, which is seen as the point at the center of the diagram in the bend in Draco the Dragon.

If you're interested in seeing the pole star when the Egyptians built the pyramids, this dim star Thuban in Draco, draw a line from the center star in the handle of the Big Dipper straight across to the bowl of the Little Dipper. Halfway in between the handle of the Big Dipper and the Bowl of the Little Dipper you'll come across a dim 4th magnitude star and that was the pole star when the Egyptian built the pyramids.

Precession is caused by the Moon's gravity pulling on Earth. You might remember that Earth isn't a perfect sphere. It bulges out at the equator and it's thinner around the poles. The Moon doesn't orbit around the Earth's equator. Instead, it's in an orbit that's tilted about 5 degrees with respect to the ecliptic, the Earth's orbit around the Sun, which is tilted about 23.5 degrees to the Earth's equator. This means the Moon is sometimes above the equatorial bulge, and sometimes it's below the equatorial bulge. As the Moon pulls on this bulge, it tries to pull the bulge up or down, it tries to tip the Earth over. But like any spinning top, if you try and tip a top over, it will wobble instead.

Another consequence of this precession, as discovered by Hipparchus, is that the equinoxes are shifting with respect to the background stars. Remember from Lecture 4 that the vernal, or spring, equinox is where the ecliptic crosses the equator; that's where the Sun goes from the Southern Hemisphere to the Northern Hemisphere about March 20th of every year. Hipparchus discovered that the equinox is slowly shifting in the sky.

Today it's located in the constellation of Pisces. But 2,500 years ago, it was located in the constellation of Aries. This is why that vernal equinox point is still to this day called "the first point of Aries." A few thousand years earlier, the vernal equinox point was in Taurus, the Bull. Some believe that this is the reason that bulls are often associated with fertility stories. That's because the vernal equinox point being in Taurus was around the spring time, and that's when we would go out and plant crops and the crops would first start growing.

Around the year 2600, the vernal equinox is going to shift from Pisces into the constellation of Aquarius. This is the basis for the lyrics "dawning of the age of Aquarius" in the 1969 song "Aquarius/Let the Sun Shine In" from the musical *Hair*.

In the second lecture, I described the system of coordinates that astronomers use to measure the positions of objects in the sky. Right ascension is the east-west position while the vernal equinox is defined as the zero point of that right ascension. Declination is the north-south position in the sky and that's measured with respect to the celestial equator.

If the vernal equinox point is shifting through the sky, that means that the starting point for right ascension is moving as well. Thus, right ascension and declination of a star change over time. Positions of objects in the star gradually change over time as the Earth wobbles around. When astronomers specify the right ascension and declination of an object, they also state the epoch, or the year with which those coordinates were measured. When you purchase a star atlas, it will list the year of the equinox used to establish the coordinate system.

All modern catalogs and star atlases use epoch 2000 coordinates, and will be using epoch 2000 coordinates for another 20 or 30 years. About every 50 years, astronomers change from one epoch to the next. Before 2000, the most common epoch system was 1950, which was commonly used in astronomy up until the 1990s.

Precession is very small, it amounts to about 1 degree shift every 72 years. Most amateur astronomers don't need to worry about coordinate systems unless they're doing precision work. There are many computer programs out there that are available, including many on the Internet, which will convert one epoch to another if you ever have to find yourself making that change.

Let's move on now to talking about the northern constellations. We'll begin with the brightest and easiest to find, Ursa Major. We've already seen the Big Dipper, the Plow, and the Wagon asterisms associated with Ursa Major. Earlier I told you that these stars represent a bear in the sky. Let's see how they become a bear.

Here's the familiar asterism of the Big Dipper. The four stars in the bowl of the Dipper form the body of the bear. Two stars form the neck of the bear. The end star forms his nose. These stars form a front leg coming down to the toes on the front paw. These stars form one back leg, coming down to two toes on the back paw. These stars form another back leg. Finally these stars form a long tail on the bear.

I think this grouping of stars looks remarkably like a stick figure drawing of a bear. Both the Greeks and the Native Americans, two civilizations widely separated by half the Earth, saw a bear in this part of the sky.

The bright stars of Ursa Major, in order from the tip of the bowl, are Dubhe, Merak, Phecda, Megrez, Alioth, Mizar, Alcor, and Alkaid. There's an easy way to remember the order of these names if you'd like to know the proper names of the stars. You can use the mnemonic "DUMP MAMAA" where the "DU" stands for Dubhe, and the two "As" at the end stand for little faint Alcor, which is a companion to Mizar, and then Alkaid which is in the end of the handle.

Fortunately, the Bayer designations are easy, since he labeled them in order along the Dipper, starting with Dubhe there Alpha, Beta, Gamma continuing on through the Dipper.

Like the Big Dipper, the Little Dipper is an asterism and it has four stars in the bowl, and three in the handle. The constellation is Ursa Minor, the Little Bear. The four stars of the bowl represent the body of the bear and the three stars in the handle represent the long tail on the bear. Polaris, the Pole Star, is the end star in the tail of the Little Dipper.

Most of the other stars in Ursa Minor are faint. In fact, other than Polaris and the two end stars in the bowl, the stars are too faint to see under light-polluted skies. That is, you may only see the two end stars in the pole in Polaris. If you can see only Polaris and the two end stars then the limiting magnitude in your sky is worse than 4. If you can see all four stars in the bowl, and all three stars in the handle, you've got very good skies because then the limiting magnitude is better than 5.

Have you figured out what's anatomically incorrect about our two bears in the sky? If you said it's the tails, you are correct. Terrestrial bears have short, stubby little tails, whereas the two celestial bears have long tails in the sky. The Greeks have a wonderful story that explains why we have two bears with long tails in the sky. I'll follow this with another tale from the Native American Micmac Indians in New Brunswick and Nova Scotia, Canada, which relates these two bears to our seasons.

Ursa Major represents the beautiful maiden Callisto in the sky. Callisto, of Arcadia, was a member of a sect of women that worshiped Diana, the goddess of the hunt. Callisto was one of Diana's favorite hunting partners.

One of the requirements to belong to this sect of women was that you remain a virgin.

Jupiter was down surveying the Earth after the ride of Phaeton and Jupiter happened to cast his eyes on beautiful Callisto and fell in love with her. Seeing beautiful Callisto, Jupiter decided to disguise himself in the guise of Artemis and trick Callisto into sleeping with him. Jupiter thought that his wife Juno would never know of the love affair and, even if she did, he thought it was worth the quarrel that would ensue. Jupiter changed himself into the likeness of Diana and approached Callisto. He embraced Callisto in an immodest way with which a goddess wouldn't embrace one of her followers.

Diana realized that this was Jupiter, and she resisted him with all her might, but she couldn't stop him. After the moon had been full nine times, Diana and her hunting party were bathing in a spring. When Callisto undressed, they saw that she was pregnant, and Diana expelled her from the group.

After her son Arcas was born, Juno, Jupiter's wife, came to the realization that the child was her husband's. Juno blamed Callisto and punished her by taking away the beauty that had made her attractive to Jupiter. Juno changed Callisto's fair skin into shaggy fur, her beautiful hands and fingers became claws, and her beautiful mouth was changed into gaping jaws. Instead of being the hunter she was the hunted. She had been turned into a bear and she had to flee from the sounds of the barking dogs.

After many years, Callisto came across a young man hunting in the woods. She realized that the young man was her son, Arcas, so she rose on her hind legs to embrace him. Thinking he was about to be attacked by a bear, Arcas drew an arrow, strung it in his bow, and was about to kill his mother when Jupiter witnessed the scene. Taking pity on poor Callisto, he threw her into the sky to become Ursa Major. He may have grabbed her by her short stubby tail and swung her around and around, stretching the tail. So that he could join his mother, Jupiter also turned Arcas into a bear and threw him into the sky, and some say that he's pictured in the sky as Ursa Minor, the Little Bear, and others believe that he is Bootes, the Bear Guardian.

That night, Juno looked in the sky and saw Jupiter's mistress and his illegitimate child on display. In her fury, she descended to the sea and asked Tethys and Oceanus to prevent the bears from ever refreshing themselves in the cool waters of the Earth. Thus, the two bears forever wander around the pole, never setting below the horizon, never dipping into the waters of Earth, which is clearly a reference to the fact that Ursa Major and Ursa Minor are circumpolar.

The Micmacs also saw a bear in this part of the sky, very similar to the Greek bear. They did not see the three stars in the handle as a tail. They thought that these three stars were hunters, Robin, Chickadee, and Moose Bird. Chickadee is represented by Mizar, which has a very dim companion next to it, Alcor, and Alcor represents the pot that Chickadee is carrying to cook the bear.

In the spring, the bear leaves his circular den, and we see the den as Corona Borealis, the Northern Crown. The hunt goes on all summer. In the fall, the bear rears up on it hind legs, and Robin shoots the bear with an arrow. The blood from the wounded bear gets all over the robin. But being a bird, Robin is able to shake the blood off of his wings, but not off of his breast, which is why robins have a red breast.

The blood that he shakes off of his wings falls to Earth and coats the leaves on the trees, turning them red in the fall. The bear dies, and is cooked in the pot carried by Chickadee. As the yellow lard bubbles out of the pot, it colors the trees yellow. After the hunt, the skeleton of the bear is high overhead in the wintertime lying on its back. In the spring, another bear leaves its den and the hunt starts again.

Now that we've heard two stories about hunting the bear, let's hunt in the bear for some interesting celestial objects that are visible. In and around Ursa Major are some of the best nearby galaxies in our sky. For example, just southwest of Alkaid, the end star in the handle of the Big Dipper, lies M51, which is the best of the nearby spiral galaxies other than Andromeda. M51 is actually in the constellation Canes Venatici, but since most observers start their hunt from M51 from Alkaid, I'm going to include it here.

M51 is a grand spiral galaxy about 31 million light years away. It's interacting with a companion galaxy whose gravity has distorted the disc of M51 and may be responsible for the beautiful spiral arms. Under dark skies, a 6- or 8-inch telescope will begin to show the spiral structure of the galaxy, but just barely. An even larger telescope will show the spiral structure more clearly.

Two other wonderful galaxies in Ursa Major are the M81, M82 pair. Through a telescope, at low magnification, both galaxies are visible in the same field at the same time. M81 is the larger of the two and it's a spiral galaxy, while M82 is an irregular galaxy and both of them are about 12 million light years away. An irregular galaxy is a galaxy that doesn't have a well-formed shape unlike the beautiful spiral structure that we saw in M51. These two galaxies are interacting with one another. Their mutual gravity has stripped material out of them, mostly gas, from both galaxies. M82 is an example of a starburst galaxy undergoing a tremendous burst of star formation.

In our Milky Way, stars are forming at a rate of about one sun-like star every year. But in M82, the star formation rate is about 10 times higher. In terms of area, Draco the Dragon is one of the largest constellations in the sky. To find Draco, draw a line from the bowl of the Little Dipper to the bright star Vega. The head of Draco is about halfway along this line. In ancient times, dragons were much more like snakes with wings rather than the larger dragons that we imagine today. A group of four stars forms the head of the dragon and its body winds along between the two Dippers. In one of his 12 labors, Hercules had to retrieve the golden apples from a tree given to Juno as a wedding present.

Draco represents the dragon that Juno was using to guard the tree. Hercules convinced the god Atlas to retrieve the apples for him by offering to relieve Atlas from his burden of holding up the heavens. When Atlas returned with the apples, he had no intention of again taking on the heavy weight of the sky. Hercules asked Atlas to hold the sky for a brief moment while he adjusted a cushion. When Atlas took the sky, Hercules left with the apples. In another version of the myth Hercules slays the dragon and Juno places the dragon in the sky.

In the fall, if you draw a line from the handle of the Big Dipper to Polaris, and keep going an equal distance on the other side of the pole, you'll reach the bright "W" shape of Cassiopeia, the queen of the kingdom of Ethiopia. Queen Cassiopeia is one of the central figures in one of the greatest celestial stories, but you'll have to wait until the next lecture to hear about the trouble that Cassiopeia caused.

If we draw a line from Alpha to Beta, the two brightest stars in Cassiopeia, and they're the two end stars in Cassiopeia, and keep going on this line about 15 degrees, we come to the constellation of Cepheus, Cassiopeia's husband and the king of the mythical of Ethiopia. Cepheus looks like a simple drawing of a house. Four stars form a square that make the structure of the house while a single star marks the peak on the roof.

One star in Cepheus is worth noting, as it's one of the most famous stars in the sky. Delta Cephei is an unstable, pulsating, variable star about 890 light years away from us. Every 5 days, 8 hours, 47 minutes, and 32 seconds it varies in brightness, it goes from magnitude 3.5 at its brightest to 4.3 at its faintest. A whole class of variable stars, called "Cepheid variable stars," are named for Delta Cephei.

Cepheid variable stars are giant stars near the ends of their lives. These stars are unstable and they vary in brightness as they pulsate. They grow huge in size and their brightness increases only to then shrink down again. They pulse with amazing regularity with periods ranging from a few days to a month or more. Cepheids are important to astronomers because Henrietta Leavitt discovered that we can use them to measure distances to other galaxies.

She discovered that the luminosity of a Cepheid-variable star, how much energy it gives off every second, depends on the length of time it takes a Cepheid-variable star to vary in brightness. Remember that the brightness of a star in our sky depends on two things. It depends on how much energy the star is giving off every second and it depends on the distance to the star. There are three things there: brightness, luminosity, and distance. If we know any two of them, we can determine the third. Usually we can measure the brightness of something in the sky, so if we can determine its luminosity we can get its distance.

With Cepheid-variable stars, we measure their brightness, we measure how long they take to vary, and that tells us their luminosity, and from that we get their distance. Because these stars are thousands of times more luminous than our Sun, we can see individual Cepheid-variable stars in whole other galaxies outside the Milky Way, including some in galaxies that are tens of millions of light years away. Determining the distances to Cepheid-variable stars was behind two of the greatest discoveries in 20th-century astronomy.

In the early part of the 20th century, it was unknown if these spiral nebula, these galaxies, were part of our Milky Way, or if they were whole other galaxies in their own right. Edwin Hubble discovered Cepheid-variable stars in the nearby Andromeda Galaxy. He showed that the distance to the Cepheids were so large, millions of light years, that the spiral galaxy had to be a separate galaxy from the Milky Way.

Less than a decade later, Hubble made his second great discovery. He found that these spiral galaxies were mostly moving away from us, and that their recession speed was related to their distance. From this he had discovered the expansion of the universe.

In fact, finding Cepheid-variable stars and measuring their periods in distant galaxies was one of the primary reasons for building the Hubble Space Telescope.

The discovery that the Andromeda nebula was a galaxy of hundreds of billions of stars separate from the Milky Way was an astounding discovery. In the next lecture we'll tour the September, October, and November sky when Andromeda is high overhead and easily visible. As we tour the sky each season we'll come back to the stars around the North Celestial Pole. In particular we will start each of the lectures with either the Big Dipper or the "W" of Cassiopeia to get the tour started, other than the winter sky when we'll start with the familiar constellation of Orion.

# The Fall Sky
## Lecture 8

We call this in astronomy star hopping, where we start on a bright star and move gradually to fainter and fainter stars to find objects.... You can't just point a telescope right at an object that you can't see, so you start at a bright star and then from a star atlas find a faint one nearby and jump to that and then finally jump to your object.

The September, October, and November sky is filled with characters from one of the great classical celestial myths. It also has the Andromeda galaxy, the greatest galaxy in our sky other than the Milky Way. It has spectacular deep sky objects and a few stars whose planetary systems represent one of the great discoveries of modern astronomy.

We can use three bright stars in Cassiopeia to find other formations, such as the Great Square of Pegasus. From there, we can find Andromeda, the chained lady. A bright line of stars down the length of Andromeda points to Perseus, who rescued Andromeda from the sea monster Cetus. The Greek myth surrounding these characters can be used as an aid to find all these constellations in the fall sky.

Almak, the end star in the left foot of Andromeda, is one of the most spectacular **visual double stars** in the sky. A visual double star is a binary star in which each of the two components is visible in a telescope. In a spectroscopic double star, the two components are so close together that even through a telescope, they are seen as one star. Astronomers use stellar spectra to differentiate these types of stars. Upsilon Andromedae is a faint star that forms the left knee of Andromeda; it is also the first star outside of our solar system discovered to have multiple planets. This discovery was also made using stellar spectra.

Another interesting object in Andromeda is the Andromeda galaxy, about 2.5 million light years away from us. It contains about 400 billion stars, and it is about 150,000 light years across. The Andromeda Galaxy is approaching the Milky Way at about 75 miles per second; the two galaxies will collide in about 3 billion years.

The constellation Andromeda can also help us find Perseus. The second-brightest star in Perseus, beta Persei or Algol, is an eclipsing binary star that has been valuable in teaching astronomers how stars live out their lives. The lifetime of a star depends on its mass. The more massive the star—the more gas it has—the higher the pressure and temperature at the center of the star. The nuclear reactions in a massive star are much faster; these stars burn up their fuel quickly and have short lives. In the case of Algol, the lower-mass star has already evolved into a giant and is dying.

The Great Square of Pegasus can be used as a test of the darkness and clarity of your sky. Under dark skies, with a **limiting magnitude** of 6 or better, you should be able to count at least 13 stars in the square. With a bit of **light pollution** and a limiting magnitude of 5, you see only 4 stars. You can also use the Great Square to find the zodiacal constellations in the fall sky.

© Comstock/Thinkstock.

**The Andromeda galaxy, at about 1.5 times the size of the Milky Way, is the largest in our local group of about 30 galaxies.**

The brightest star in Pisces Austrinus (the Southern Fish) is Fomalhaut, a fairly typical A3 hot blue star about 25 light years away. Observations in 1983 showed that Fomalhaut is surrounded by a warm disk of dust. This was tantalizing evidence that our solar system may not be alone, because planets likely form from these disks of dust as the dust clumps together. In 2008, images taken by the Hubble Space Telescope clearly showed not only the ring of dust but also a planet, now called Fomalhaut b. ■

## Important Terms

**light pollution**: The illumination of the night sky by stray light from human activities.

**limiting magnitude**: An estimate of the apparent magnitude of the faintest star that can be seen high overhead. An excellent dark sky site will have a limiting magnitude of 6, but a typical urban area with some light pollution will have a limiting magnitude of 4.

**visual double star**: A binary star in which each of the two components is visible in a telescope.

## Suggested Reading

Ridpath, *Norton's Star Atlas*, chap. 1 and charts 3–4.

Schaaf, *A Year of the Stars*, chaps. 12–14.

Dickinson, *Nightwatch*, charts 11–14, 19–20.

## Questions to Consider

1. What is the limiting magnitude from your observing location? Use the constellation of Pegasus, the stars of Ursa Minor, or a star atlas that lists the magnitudes of the stars to determine the limiting magnitude from your most common observing location on a good night. What is the faintest star you can see with the naked eye?

2. When is the next time that the star Algol is at minimum brighness? Use the Internet or an astronomy magazine to determine when the variable star Algol (an eclipsing binary) will again be at minimum brightness. Observe the star and compare it to nearby stars a few hours before or after it reaches minimum light.

3. How many extrasolar planets are currently known? The number increases every year and can be found by searching the Internet. How many systems have more than one planet? Which planet is most like the Earth? Can you see the star it orbits in the night sky?

# The Fall Sky
## Lecture 8—Transcript

The September, October, and November sky is filled with characters from one of the greatest of the classical celestial myths. It is also has the Andromeda Galaxy, the greatest galaxy in our sky other than the Milky Way. It has spectacular deep sky objects, and a few stars whose planetary systems represent one of the great discoveries of modern astronomy.

Let's begin our tour of the sky with a star map or planisphere set for 10:00 p.m. on October 15th. The sky will appear just like this at midnight in mid-September, and at 8:00 p.m. in mid-October. The first thing we want to do is find the distinctive "W" shape of the constellation Cassiopeia high overhead, near the center of the window. You'll remember that Cassiopeia was the queen of the mythical kingdom of Ethiopia.

We want to use the bright stars on the right side of Cassiopeia, the two end stars, for finding other constellations. The three bright stars in Cassiopeia that we'll use are Alpha, Schedar, which is at the bottom of the "W," Beta, which is at the top of the "W," and Gamma, which is the star in the middle. Draw a line in the sky from Alpha through Beta, the end star, and you'll come to a square of faint stars about 15 degrees away. These are the stars of Cassiopeia's husband, Cepheus, the king of the mythical kingdom of Ethiopia.

Let's use Cassiopeia to find the Great Square of Pegasus. Draw a line from Gamma, the star in the middle of the "W," through Alpha. Go more than twice as far as you went to find Cepheus in the sky and you'll come to a big square of stars about 35 degrees away from Cassiopeia. These stars form the Great Square of Pegasus. Let's use the Great Square of Pegasus to find the constellation Andromeda.

We start in the northeast corner of the square, that's the corner closest to Cassiopeia. That marks the head of Andromeda, the chained lady. We can follow a bright line of stars down the length of Andromeda, and they point to the great hero Perseus in the sky who's right at Andromeda's feet and rescues Andromeda from Cetus the Sea Monster, or the Whale.

Let's go back to Pegasus for a second and use Pegasus to find the Sea Monster. To find Cetus, draw a line from the two eastern stars in the Great Square. These are the two stars on the left-hand side, and follow them towards the south and you'll arrive at a moderately bright star and it's relatively isolated, all by itself, named Diphda, which marks the tail of Cetus the Sea Monster.

All of these characters are involved in the story of the rescue of Andromeda. I think it's the greatest of the classical celestial myths. Andromeda was the daughter of Cassiopeia, queen of Ethiopia, and Cepheus, the king of Ethiopia. This wasn't the modern day country of Ethiopia but rather a mythical kingdom said to be on the Levantine coast near the modern day Israel.

Cassiopeia was a very vain woman, and she once bragged that she was fairer than the 50 beautiful Nereids, the water nymphs. The Nereids complained to Poseidon, the God of the Seas, about Cassiopeia bragging that she was more beautiful than they were. Neither Poseidon nor the Nereids are represented in the sky. But angry Poseidon shook the brine of the sea and created a flood and a sea monster by the name of "Cetus." You'll remember that Cetus is far south in the sky at this time of the year. Cetus went to the shores of Ethiopia where it terrorized the population.

In a desperate effort to stop the sea monster, Cassiopeia's husband, King Cepheus, who's located next to her in the sky, consulted the Oracle of Ammon. The oracle said that in order to save the kingdom, Cepheus and Cassiopeia would have to sacrifice their beautiful daughter Andromeda to the sea monster. Andromeda was led to the sea and was chained to the rocks to await her fate.

Next to Andromeda in the sky we find the great hero Perseus. Perseus happened to be returning from slaying the Gorgon Medusa. You'll remember that Medusa was a witchlike character with snakes for hair. She had wings and claws on her feet. Any man that should happen to look on the face of Medusa was instantly turned to stone. To help him in his quest to kill

Medusa, Perseus was given winged sandals by Mercury and a sword and a polished shield by Minerva.

Minerva instructed Perseus not to look directly at Medusa, but to look at her reflection in the shield. Perseus slay the Gorgon by cutting off her head, and from the blood of Medusa sprung the flying horse Pegasus. Pegasus is located at the head of Andromeda in this part of the sky. In ancient Greek myths, Perseus returns home on the winged sandals. But in European Renaissance and modern imagery, he's often pictured flying home on Pegasus, the Flying Horse.

Perseus carries with him the head of Medusa in a bag. The star Algol in Perseus represents the head of Medusa. Upon reaching the kingdom of Ethiopia, Perseus looks down and he spies the beautiful Andromeda chained to the rocks below. She was so pale and so fair and motionless that he thought at first she was a statue. But he that she was crying and he could see her hair blowing in the wind and realized that it was a woman.

Perseus landed and asked Andromeda why she was there and where she was from. At first she was too shy to answer, but she relented and she told Perseus the story of her mother's vanity and her predicament. At just that moment, the sea monster Cetus rose out of the sea. With no time to spare, Perseus asked Cassiopeia and Cepheus if he could have Andromeda's hand in marriage if he could save her. Having no other choice, King Cepheus agreed.

Perseus took up a position next to Andromeda and told her to shield her eyes and he removed the head of Medusa from its bag. Cetus was instantly turned to stone when it gazed upon the face of Medusa. Later, at the wedding, Cepheus' brother Phineus demanded Andromeda, as she had been promised to him earlier. Perseus used the head of Medusa to turn Phineus and his men into stone as well.

Cassiopeia wasn't punished for claiming that she was more beautiful than the Nereids and so Cassiopeia was placed in the sky and she is forced to forever circle the pole, often having to hang upside down. After her death Andromeda was placed in the heavens by Minerva to be near Perseus, and all of the constellations related by this one story, Cassiopeia, her husband Cepheus, their beautiful daughter Andromeda, Perseus riding Pegasus the Flying Horse to save her from Cetus the Sea Monster, are all in the September, October, November sky.

This is one of the reasons that I tell you this story because if you can remember it you can use it as an aid to find all of these constellations in the fall sky. Let's turn our attention now to one of the constellations. We'll turn our attention to Andromeda to begin with.

It's easiest to find Andromeda by starting with the Great Square of Pegasus. To see the Chained Lady, you must imagine a stick figure drawing of a woman with her arms outstretched, chained to a rock wall. Alpheratz, or Sirrah, which is the northeast star in the Great Square, the one on the upper left which is closest to Cassiopeia, forms the head of Andromeda.

Her body is represented by a line of stars that stretch from Alpheratz to Mirach. Her legs are formed by two gently-curving strings of stars and her outstretched arms are chained to the rock and they're represented by two faint lines of stars coming from her upper body. Along the length of Andromeda there are three main stars. Alpheratz, which also has the name "Sirrah," is the brightest star in Andromeda so it's also known as "Alpha Andromedae." The name is interesting. It's actually derived from corruptions of the Arabic for horse, "*alpheratz*" or "*sirrah alpheratz*," which stands for the horse's navel.

You can see that this star not only represents the head of Andromeda, but also used to represent the belly of the horse. In fact Alpheratz is one of only two stars in the sky that carried two Bayer designations. It was not only the head of Andromeda, Alpha Andromedae, the brightest star in Andromeda, but it's also the shoulder of Pegasus the Flying Horse. It was also known as Delta Pegasi. The other star that has two Bayer designations is the star Almak in Taurus and Orion.

When the boundaries of the constellations were set by the International Astronomical Union for the 88 modern constellations in 1930, Alpheratz was placed in Andromeda. The name Delta Pegasi is no longer used.

Let's now look at some of the bright stars in Andromeda. Almak is the end star in the foot of Andromeda. It's her left foot. It's one of the most spectacular visual double stars in the sky. There are two kinds of double stars that we'll be interested in observing. A visual double star is a star where each of the two components are visible in a telescope. You can actually see the two stars next to one another. In some cases as these two stars orbit around one another, you can see the motion of the two stars orbiting one another after only a few years of watching.

Another kind of binary star that we'll talk about in this course is spectroscopic binary stars. In a spectroscopic binary star the two stars are so close together that we can't separate them. Even through a large telescope they look like a single star. But we can tell that these stars are double by using a spectroscope to break the light from the star into its component colors, just as a prism breaks white light into a rainbow. Dark lines in the spectrum of the star come from elements absorbing particular colors of light. For example look at these stellar spectra here. We see absorption lines, dark lines from hydrogen atoms, sodium atoms, calcium atoms, and many others.

Astronomers can use spectra like these to tell what stars are made of, or to differentiate the different types of stars. We use this to classify stars into O, B, A, F, G, K, and M stars and we also use these dark lines to determine if a star is moving towards us or away from us. If a star is moving toward us, the dark lines are shifted very slightly toward the blue end of the spectrum, while if the star is moving away from us, the lines are shifted very slightly to the red.

When astronomers monitor a star that we think is a spectroscopic binary star we look at these dark lines. If it's a spectroscopic binary we'll see the lines shifting back and forth as the stars orbit one another and the star comes towards us and away from us and towards us and away from us.

Almak is a visual binary star. That means we can see the two components and even a small telescope shows a magnitude 2.3 golden yellow star with a magnitude 5.4 beautiful blue companion about 10 arc seconds away. It's the colors that make this such a beautiful star. The yellow and the blue, or some people have described it as topaz and aquamarine or orange and emerald. The fainter blue companion is itself a visual binary star, but the separation is only 0.3 arc seconds.

To see both of these components of the pair requires a large telescope under excellent seeing conditions and high power. One of those stars happens to be a spectroscopic double star meaning that the two stars are so close together that we can't see them even in a large telescope. That makes Almak a quadruple star system.

Upsilon Andromedae is a faint, $4^{th}$ magnitude star forming the left knee of Andromeda. It's a rather normal F star, about 44 light years away, but what's interesting about it is not the star itself, but what's going around the star. One of the greatest recent developments in astronomy in the last few decades has been the discovery of hundreds of planets around sun-like stars. Upsilon Andromedae was the first star outside of our solar system to have discovered around it multiple planets.

There are three confirmed planets in orbit around Upsilon Andromedae. These planets can't be seen in a telescope, but it's worth keeping this in mind when you go out and look at Andromeda in the sky and see Upsilon Andromedae and realize that there are three planets around that star. Astronomers have detected the planets the same way that we detect spectroscopic binary stars. We do that by breaking the light from the stars into their component colors.

As the planet orbits the star the star wobbles back and forth in response to the subtle gravity from the planet. Like in a spectroscopic binary star we detect the shifting of the dark lines in the star as it comes towards us and away from us and towards us and away from us. But because the planet orbiting around it is so much less massive than a star, the resulting wobble of the star is very small. It's much, much less than in a spectroscopic binary and it's very difficult to see.

We've been able to do that and with it we've discovered three planets going around Upsilon Andromedae and they're all massive, gas giants. They have

0.7, 2.0, and 4.0 times the mass of Jupiter. Remember the mass of Jupiter is 318 times the mass of the Earth.

What was most surprising though about these planets was that the inner one takes only 4.6 days to go around the parent star, while the other two take 242 days and 3.5 years. Remember Mercury in our solar system. Mercury takes only 88 days to orbit our Sun. The inner planet in Upsilon Andromedae, going around every 4.6 days, must be many times closer to its parent star than Mercury is to our Sun. That planet must be incredibly hot.

The next most interesting object in the constellation Andromeda is not a star but actually one of our deep sky objects, one of our nebula. It's the best galaxy in the sky other than the Milky Way Galaxy. The Andromeda Galaxy is about 2.5 million light years away from us. It's the farthest object that you can see with the naked eye. It's the closest example of large spiral galaxy to our Milky Way.

It contains about 400 billion stars, which is about twice as many stars as our Milky Way contains, but our Milky Way Galaxy turns out to be as massive or slightly more massive. It's tipped not quite edge-on, but about 77 degrees to our line of sight, so we're viewing it almost edge-on. And it's about 150,000 light years across, which is 1.5 times the size of our Milky Way.

The Andromeda Galaxy is the largest galaxy in our local group of galaxies. Our local group, our neighbors in the universe, and they consist of, well there's the Milky Way and M31 Andromeda, another spiral galaxy M33, and then about 30 smaller galaxies in the local group as well. The Andromeda Galaxy is approaching the Milky Way at about 75 miles per second. That's about 120 kilometers per second. The two galaxies will actually collide with one another in about three billion years. This was the first galaxy that was known outside of our Milky Way.

Edwin Hubble had discovered Cepheid-variable stars in the Andromeda Galaxy in 1924. He showed that it was so far away that it wasn't part of our Milky Way, but it was a whole galaxy in its own right.

Observing the Andromeda Galaxy is an easy thing to do. It's barely visible with the naked eye, even under some skies with light pollution. To find M31, we want to start with the head of Andromeda, that star Alpheratz. We jump down two stars along the body of Andromeda to Mirach and then we come across two stars to Nu Andromedae.

We call this in astronomy "star hopping" where we start on a bright star and move gradually to fainter and fainter stars to find objects. It's something you'll do quite often when you're looking for objects in the sky. You can't just point a telescope right at an object that you can't see, so you start at a bright star and then from a star atlas find a faint one nearby and jump to that and then finally jump to your object.

Here we're starting at Alpheratz, coming down two stars to Mirach, which is a little bit fainter, across two stars to Nu Andromedae, which is fainter still, and then finally 1.3 degrees away to the northwest of Nu Andromedae, we'll see the Andromeda galaxy. It's easiest to see using averted vision. Don't stare right at it, in fact it's actually useful I've found to stare close to Nu Andromedae and using your averted vision, out of your peripheral vision there, you will see a faint fuzzy cloud in the sky. This faint fuzzy cloud has probably been known since antiquity because it's visible with the naked eye.

Through a pair of binoculars under very dark skies, you'll get a better appreciation for the size of the Andromeda galaxy. Its total length can be traced over 4 to 5 degrees, which is about 8 to 10 times the diameter of our full Moon. So this galaxy is very large in the sky.

It's so large that when you observe M31 with a telescope you'll need to use the lowest power eyepiece that has the widest possible field of view. Even in that lowest power, the galaxy is likely to completely fill the field of view. The center of M31 through a telescope will appear as a yellow-white ball just a few arcminutes across. Under high magnification you can see the bright, star-like nucleus of the galaxy.

As you tour around the galaxy, you may start to see dark dust lanes where giant clouds of gas and dust block our view of the stars behind. These giant dust lanes are obvious in the photographs, but they're even obvious when you look through a telescope. Under dark skies you will see that there's a dark lane cutting across the galaxy.

Southwest of the nucleus, you may even be able to make out NGC 206. It's a somewhat brighter patch in the disk of the galaxy. It's a bright cloud of stars and it's one of the largest star forming regions in the local group. Keep in mind that when we look at Andromeda we're looking at a galaxy just like the Milky Way filled with clusters of stars, star forming regions, hundreds of billions of individual stars, but the galaxy is so far away that we can't make out the individual ones. NGC 206 is one bright star forming region that you can make out in Andromeda.

Andromeda has two moderately bright companion galaxies, M-32 and M-110, that are visible in a telescope. Both of these are dwarf elliptical galaxies, meaning that they're elliptical galaxies that contain low mass, cool, red stars. There are no young, hot, blue stars in elliptical galaxies because they've used up the gas that you would need for star formation long ago and all the hot, blue stars have died. The only things we see left are the cool, red and yellow stars.

These two galaxies next to Andromeda are dwarf ellipticals, which means they're smaller than regular or giant elliptical galaxies that we'll see later. These two galaxies typically contain between a few and 10 billion stars.

Just south of Andromeda is the constellation of Triangulum. This is where we're going to find another one of the local group members, that spiral galaxy M33. As its name implies, Triangulum is formed by three moderately bright stars. It's home to M33, which is the smallest of the three spiral galaxies in our Local Group, with the other two being the Milky Way and the Andromeda Galaxy. M33 is just slightly farther away than Andromeda, about 2.6 million light years from us. Unlike Andromeda, which we're viewing almost edge-on, M33 is viewed almost face-on, which is how it got its nickname, "The Pinwheel Galaxy."

Given how close it is to us, and its face-on orientation, you might think that it would be one of the most spectacular objects in the sky. But its light is spread out over such a large area, about 1.2 by 0.7 degrees, that it's a very difficult naked eye object that requires absolutely dark skies. M-33 is visible in binoculars with their wide field and their low magnification, but it can be challenging in a telescope.

If you want to see it in a telescope you need to use your lowest magnification eyepiece that has the widest field of view. M-33 itself, as a galaxy, is somewhat smaller than the Milky Way, about half the diameter and about 1/7 the mass. It's visible here in Triangulum right next to Andromeda in the sky.

We can use Andromeda to find Perseus in the sky. Remember that we start with the head of Andromeda, Alpheratz, and follow the body of Andromeda down through Mirach past her foot Almak and continue on. The next bright star that you reach is Mirfak, which is also known as Algenib or Alpha Persei, the brightest star in Perseus.

The pattern of Perseus is distinctive. I think it looks like a large curved "T" in the sky, but it's difficult to fit the stick figure drawing of a hero to these stars. The stars that are northwest of Mirfak form the body and the head of Perseus and two bent lines of stars represent his legs. The second brightest star, Beta Persei or Algol, represents the head of Medusa.

Alpha Persei, or Mirfak or Algenib, is an F class supergiant star. Surrounding it is the Alpha Persei Cluster. You can see the Alpha Persei Cluster with the naked eye or with a pair of binoculars and it's a cluster of stars surrounding Algenib or Mirfak. The cluster, including Mirfak, is about 600 light years away from us.

Algol, on the other hand, Beta Persei, is one of the most interesting stars in the sky. It represents the head of Medusa and its name comes from the Arabic *ra's al-ghūl,* "the Demon's Head." It's related to our word "ghoul." It was Ptolemy, around 150 A.D., who designated this star as the head of Medusa. In 1667, it was first noted that the star is a variable. Every 2 days and 20 hours, 48 minutes, and 58 seconds, the stars, it's a binary star system, go into eclipse. During an eclipse, the star dims from magnitude 2.1 to magnitude 3.4. It takes about five hours for the stars to dim and another five hours for the stars to recover.

Today we know that Algol is an eclipsing binary star. As you can see in this video, the binary star system consists of a hotter, blue B8 star that goes behind and is then eclipsed by a cooler, yellow-orange K star. In every orbit, the K star passes in front of the brighter B star blocking nearly 80 percent of its light. When that light is blocked the star dims down. The dimming is easily visible with the naked eye.

Every time you go out and look at Perseus, compare the brightness of Algol to the nearby stars, especially Rho Persei. Most of the time, Algol is significantly brighter than Rho, but during an eclipse, it will be nearly as faint or fainter. If you'd like to go out and plan to see an eclipse the times of the eclipses are well known, and they're published in astronomy magazines and they're available on the Internet. Just search for "Minima of Algol" to find out when it will go into eclipse.

Algol has been very valuable for telling astronomers about how stars live out their lives. The lifetime of a star depends on its mass. The more massive the star, the more gas it has, the higher the pressure and the temperature down at the center of the star. The nuclear reactions in a massive star are much faster and these massive stars burn their fuel up more quickly and they have short lives.

Lower mass stars burn their fuel more slowly. The temperatures and pressures at the center of the star aren't as high. The nuclear reactions are going slower and so these low mass stars live longer lives. So Algol is an interesting star. The lower mass star, the K mass star, which is about 0.8 times the mass of our Sun, has clearly evolved into a giant star. It's already dying. The more massive star, which should lead the shorter life, is about 3.7 times the mass of our Sun. It's odd that the low mass star is dying before the high mass star.

Today we know how the paradox is resolved. It's resolved by the fact that the low mass star was originally the higher of the two masses. It was the more massive star, and it lived out its life, and when it died it puffed itself up into a red giant. It got so big, in fact, that its outer atmosphere was so close to its companion star that the companion star started pulling gas off of the more massive of the two stars. Today enough gas has been lost from that initial massive star that it's actually the less massive of the two. It's this transfer of mass from one star to the other that explains why the lower mass star is dying before the higher mass star.

With the naked eye under moderately bright skies, you might notice a faint fuzzy patch halfway between the central star of Cassiopeia and Mirfak and Perseus. The easiest way to see it is just scan your eye from Cassiopeia to Perseus and back again, and under dark skies you'll see this faint fuzzy patch. This patch is the Double Cluster. It's also known as "H" and "chi Persei," or it's cataloged as NGC 869 and NGC 884. It's two open clusters of stars, side by side. The Double Cluster, like the Andromeda Galaxy, is visible with the naked eye so it's probably been known since antiquity. We certainly know that Hipparchus recorded it in his star catalog around 150 B.C.

Through a small telescope, the double cluster is one of the finest sights in the sky. In a small telescope, you'll see dozens of bright stars in each of the two clusters. Since the two clusters are only a degree apart, and each is about 30 arcminutes wide, a low power, wide field eyepiece will fit both stars in the field of view at the same time. Both clusters are about 7,300 light years from Earth, and each them is about 70 light years across. We know their ages, they're both about 12 million years old. Since they're located together in the sky and they both have the same age, it's like that both clusters formed at the same time from the same giant star-forming region.

Let's turn our attention now to Pegasus, the Flying Horse. You'll remember that Pegasus is adjacent to Andromeda in the sky. Four bright stars, Algenib, Markab, Scheat, and Alpheratz form the Great Square of Pegasus. We'll see in a minute that this Great Square is useful for finding other constellations. The four stars in the Great Square of Pegasus form the body of the horse, and in the Great Square of Pegasus, the horse that we see in the sky, is only the front half of the horse.

If we go to the lower right-hand corner of the Great Square we see a line of stars stretching to the southwest to the lower right from Markab, which forms the neck of the horse. A little above this we have the star Enif, which is derived from the Arabic word "*anf*," for nose. It marks the nose of the horse. Two lines of stars heading west out of Scheat represent the front legs of the horse. We have a stick figure drawing of the front half of a horse in this part of the sky.

The Great Square can be used as a test of the darkness and clarity of your sky. Under dark skies, with a limiting magnitude of 6 or better, you should be able to count at least 13 stars in the square. With a bit of light pollution, and a limiting magnitude of 5, you'll only see four stars. If you don't see any stars inside the square, your limiting magnitude is 4 or worse.

The last constellation we'll discuss is Pisces Austrinus, the Southern Fish. If you draw a line from Scheat through Markab, and continue another 40 degrees to the south, you come to the brilliant, 1st magnitude star Fomalhaut. It's the brightest star in Pisces Austrinus, the Southern Fish, which is really just a faint circle of stars that has brilliant Fomalhaut on the eastern side. It's name actually means the mouth of the Southern Fish. Fomalhaut's a fairly typical A3 hot, blue star about 25 light years away.

Observations with an infrared astronomical satellite in 1983 showed that Fomalhaut is surrounded by a warm disk of dust. This was tantalizing evidence that our solar system may not be alone, since planets likely form from these disks of dust as the dust clumps together. But at the time, no telescope had the ability to see the planets or to detect any effect they might have on this disk of dust.

But in 2008, images taken by the Hubble Space Telescope clearly showed not only the ring of dust, but they also showed a planet now called "Fomalhaut B." Along with another star that was announced at the same time, this was the first ever visible light image of a planet outside our solar system around another star. The planet that orbits Fomalhaut goes around every 872 years at a distance of 115 Astronomical Units. Though the disk and the planets aren't visible when you look up at Fomalhaut, keep them in mind, because it's a bright star that's easily visible with the naked eye.

Though all these constellations are highest in the fall, remember that you can see them at other times of the year. In the summer months they're high up in the sky as the summer constellations are setting around 2:00 to 4:00 in the morning. In the winter, you can catch a glimpse of these fall constellations just before the Sun sets. In the next lecture, we'll see the constellations of the winter sky.

# The Winter Sky

## Lecture 9

It's remarkable that many cultures, widely separated, believe that there should be seven stars in the Pleiades, but we can only see six. This has led to the idea of a "lost Pleiad," that is, that one of the stars must have been substantially brighter in the past and is dimmer today. Some of the stars in the Pleiades are variable, but none of them varies significantly enough to explain this universal idea of a missing Pleiad.

The winter sky has in it the story of the birth, life, and death of the twin brothers Castor and Pollux. So, too, at this time of year, the stars that are visible tell us about the birth, lives, and deaths of stars.

For the winter sky, we will use the constellation Orion as our starting point. Orion is accompanied by his two hunting companions, Canis Major and Canis Minor. Sirius, the brightest star in our night sky, is in Canis Major. This star, together with Betelgeuse in Orion and Procyon in Canis Minor, forms an **asterism** called the Winter Triangle. We can use Orion to find Taurus; the Pleiades, or Seven Sisters; Gemini; and Auriga the Charioteer.

The Pleiades, an open cluster about 435 light-years from Earth in the constellation Taurus.

The most famous star in Orion is Betelgeuse, an M-class red **supergiant**. It is the largest star of any kind within a few thousand light years of the Sun, and its size allows us to see its surface with telescopes. Stars generate energy by fusing hydrogen atoms into helium atoms in their cores. A star starts with a limited amount of hydrogen, and when it runs out, the star will begin to die and puff itself up into a **red giant**. Betelgeuse is one of these red giant stars.

In the sword of Orion, we find M42, the Orion Nebula. An immense amount of detail can be seen in the swirling, greenish-white nebula, especially under dark skies. At the center of the nebula is a cluster of four newborn stars called the Trapezium cluster. Also near the constellation Orion is the Horsehead Nebula. Sirius (the Dog Star) in Canis Major has a companion, Sirius B (the Pup), that is a **white dwarf star**. Such stars are the burned-out cores of low- to moderate-mass stars.

**Betelgeuse is one of these red giant stars, and it's near the end of its life. … It's far enough away that it doesn't present any special threat to Earth, but nevertheless, it will be spectacular when it goes.**

Following the belt stars of Orion up and to the right, we come to Taurus the Bull, which is the home of two **open clusters** of stars. Star clusters come in two types: open and globular clusters. Open clusters contain about 50 to 1000 stars and are primarily found along the plane of the Milky Way. **Globular clusters** contain between 10,000 to 1 million stars, and they are usually found surrounding the Milky Way in all directions in a giant halo but concentrated toward the center of our galaxy. Clusters are where astronomers study how stars are born, how they live, and how they die.

The Pleiades have been important to mark the times for planting and harvest throughout the world. Within recent times, the Pleiades seem to have wandered into a thin cloud of dust. As the dust scatters the blue light of the stars, it creates a faint blue reflection nebula that can be seen around the brighter members under dark skies.

Another open cluster near Taurus is the Hyades, used by astronomers to calibrate the celestial distance ladder. As Earth goes around the Sun, we see the nearby stars from two different perspectives. The size of the shift in the position of a star can tell us the distance to the star, but beyond about 300 light years, this **parallax** shift is too small to measure accurately. Astronomers must use brightness and **luminosity** information from nearby stars to determine distances to farther stars. The Hyades serves as an important cluster for calibrating these measurements. ■

**asterism**: An easily recognizable pattern of stars in the sky, such as the Winter Triangle.

**globular cluster**: Region typically containing 10,000 to 1 million old stars.

**luminosity**: A measure of how much light or energy a star gives off every second. Luminosity is an intrinsic property of the star that is related to its size and surface temperature.

**open cluster**: Region typically containing 50 to 1000 young, hot, blue stars.

**parallax**: Change in the apparent position of an object when viewed from two different positions. In astronomy, this phenomenon is used to calculate the distance of stars from the Earth.

**red giant**: A low- to moderate-mass star in the last stages of its lifetime; such a star has fused all the hydrogen atoms at its core into helium atoms.

**supergiant**: A massive star of great luminosity.

**white dwarf star**: The burned-out core of a low- to moderate-mass star.

## Suggested Reading

Dickinson, *Nightwatch*, charts 15–18.

Monroe and Willamson, *They Dance in the Sky*.

Ridpath, *Norton's Star Atlas*, charts 5–8.

Schaaf, *A Year of the Stars*, chaps. 3–5.

1. Why is it rare for us to be able to resolve features on the surface of another star?

2. If the Orion Nebula is one of the best nearby star-forming regions, why do we see only a few dozen stars in and around the nebula?

3. Why is Sirius the brightest star in our night sky? How does its luminosity compare to the Sun? How does its distance compare to the distances of the other bright stars?

4. Open clusters are typically 30 light years across while globular clusters can be 50-300 light years across. If globular clusters are larger, why do they usually appear smaller in a telescope?

5. Why do globular clusters not have any blue stars in them?

6. Why is there an upper limit to how far astronomers can use parallax to measure the distances to stars?

# The Winter Sky
## Lecture 9—Transcript

The winter sky of December, January, and February has in it the story of the birth, life, and death of the twin brothers Castor and Pollux. So, too, this time of the year the stars that are visible tell us about the birth, lives, and deaths of stars. Let's begin with a tour of the winter sky.

In the fall sky we were talking about the stars of Cassiopeia and used those to find Pegasus the Flying Horse. But now that the constellations have moved and the seasons have changed the celestial sphere has rotated so that what's visible this time of the year is the constellation Orion. We'll use Orion as our starting point to find the other constellations. If you have a planisphere you'll want to set the planisphere for January 15th at 10:00 p.m. to see the stars like this map.

In Lecture 1 I mentioned that Orion is the brightest and easiest to find of the winter constellations. Orion forms a stick figure drawing of a man and the most distinctive part of the constellation is the three stars that form the belt of Orion. Two stars, Bellatrix and Betelgeuse, form his shoulders, two stars form his knees, Saiph and Rigel. The belt is tilted because there's a heavy sword hanging off one side, a small group of stars above his shoulders represents Orion's head. His right arm is holding a club over his head while in his left hand he's holding a shield.

Like all hunters in ancient times Orion is accompanied by his two hunting companions, Canis Major and Canis Minor, the Large and Small Hunting Dogs. Let's use Orion to find his two companions. First start with the belt stars of Orion, follow the belt stars of Orion southeast to the lower left about 20 degrees, which is the width of your outstretched hand, and we come to the bright star Sirius in Canis Major, the Large Hunting Dog. Sirius is the brightest star in our night sky. Though some of the planets, like Venus and Mars and Jupiter can appear brighter, they're never in this part of the sky, which is below Orion's belt, because the ecliptic, the path of the planets through the sky, passes well north of Orion.

Let's use Orion to find his other hunting dog companion. Start with the two shoulder stars, draw a line from Bellatrix to Betelgeuse and go another 30 degrees and you'll come to the bright star Procyon in Canis Minor. Betelgeuse, Procyon, and Sirius form a large triangle, an asterism called the "Winter Triangle," which is a good way to get yourself oriented in the winter sky.

Now let's use Orion to find one of the bright zodiacal constellations this time of the year. Let's go back to the belt stars and this time if we follow the belt stars up and to the right we come about 20 degrees, we come to the distinctive "V"-shaped face of Taurus the Bull with its red eye, Aldebaran. About 15 degrees further west we come to the Pleiades, or the Seven Sisters, which is a tight grouping of six bright stars and we'll see later it's an historically important sign for farmers.

Now let's use Orion to find another bright zodiacal constellation. Starting with the belt of Orion, let's draw a line that goes up through Betelgeuse, north about 40 degrees, and you reach two bright stars Castor and Pollux in Gemini the Twins. Finally, we'll use Orion, and go due north out of Orion about 45 degrees, and come to the brilliant star Capella in Auriga, the Charioteer.

Now that we've taken a quick tour of the winter sky, let's look at each constellation in more detail. Since some of the planets are moving through the ecliptic, I can't tell you which constellation they'll be in this time of the year. But we do know that they'll be somewhere along the ecliptic. Look back at your star chart or your planisphere and find the ecliptic running through Taurus, Gemini, and Cancer. If there's a bright planet in the winter sky, it will be near this line.

Since we used Orion to find the other constellations let's begin with him. When we look at Orion, it's the bright stars that stand out. Not only are they impressive to look at with the naked eye, but once we learn more about them, they're even more impressive.

Probably the most famous star in Orion is his upper right-hand shoulder, Betelgeuse. In the first lecture I told you the story of how Betelgeuse got its name. Physically it's one of the most interesting stars in the sky. It's a class M, red supergiant star about 640 light years away from us. What makes Betelgeuse special is it's the largest star of any kind within a few thousand light years of the Sun. If we place Betelgeuse where our Sun is, its atmosphere would swallow up Mercury, Venus, Earth, Mars, the asteroid belt, and would go all the way out to the orbit of Jupiter. Its radius is about 5 astronomical units.

Most stars appear to us as a tiny little point of light. Even in the largest telescopes or even the Hubble Space Telescope stars are too small to show any details. Only a couple of the largest stars have ever had their surfaces imaged by telescopes, and Betelgeuse is one of those.

Here's an image taken by a special set of telescopes where multiple telescopes are combined together to act like one giant telescope. With this telescope we're able to see the surface of Betelgeuse. What we can see on the surface of Betelgeuse are some hot spots on the surface. The size of Betelgeuse in this image is only 0.04, four-hundredths of an arcsecond across. To give you an idea of how small Betelgeuse appears, that's the size of a small coin, like a U.S. nickel, as seen from 68 miles, or 109 km, away. That said, Betelgeuse is still the largest star in our sky after our Sun.

Stars generate energy by fusing hydrogen atoms into helium atoms in their cores. A star starts with a limited amount of hydrogen, and when it runs out, the star will begin to die and puff itself up into a red giant star. Betelgeuse is one of these red giant stars and it's near the end of its life. It started life as a massive O-type star, a blue star about 20 times the mass of our Sun. Betelgeuse will end its life in a titanic supernova explosion. It's far enough away that it doesn't present any special threat to Earth, but nevertheless it will be spectacular when it goes. For a few months, it will shine in our night sky brighter than the full Moon and will be visible in full daylight.

Of all the stars that are easily visible with the naked eye, my belief is that Betelgeuse is the most likely to go supernova. But because we don't know what's going on deep down in the star, all I can tell you is that it will probably go supernova in the next few million years. Rigel, you'll remember, is the star that forms the foot of Orion and its name is *rijl al Jawza* from the Arabic, which means "foot of al Jawza." Like Betelgeuse is a supergiant star, this is a blue supergiant star about 860 light years away.

In the sword of Orion, hanging off of Orion's belt, we can find M42, the Orion Nebula. It looks like a fuzzy star with the naked eye, but through a telescope the Orion Nebula becomes one of the most spectacular objects in the sky. This drawing shows very much how it appears through an eyepiece in a telescope. It looks like a fuzzy greenish-white patch of cloud. In the brightest part of the nebula, right at the center, you'll clearly be able to see a cluster of four newborn stars, called the "Trapezium Cluster."

You'll see an immense amount of detail in the swirling, greenish-white nebula especially under dark skies. You should be able to follow the swirls and the individual streamers of gas for quite some distance from the center of the cluster.

Now here's a long exposure of the Orion Nebula taken with a camera. What you'll notice in the photograph of the Orion Nebula is that the Nebula looks much brighter and that's because in photographs we can expose the film for much longer than your eye can. In fact in a photograph like this we can expose the film to the sky for many minutes or in some cases even hours. In addition, a camera is much more sensitive to red light than your eye is. For this reason many of the nebula in the sky look very different than those wonderful astronomical photos that you're used to seeing. At a distance of 1530 light years, the Orion Nebula is the closest large star-forming region to our Sun.

The Trapezium stars, right at the center of the cluster, are all less than a million years old and they were born from this giant cloud of gas and dust. Ultraviolet light from the Trapezium stars are exciting the gas in the surrounding cloud to glow. The brightest of the four Trapezium stars is an O star that's about 50 times the mass of our Sun. It's the one responsible for generating most of the ultraviolet light that excites the surrounding gas to glow.

The visible Orion Nebula, what you can see in a telescope, is actually just a blister on the surface of a much larger cloud of gas and dust. Since most of the cloud is cold, and distant from hot stars that can light it up or heat it up, it's invisible in the kind of light that we can see. But the little tiny grains of dust in that cloud are warm, and they're glowing brightly in the infrared portion of the spectrum.

If you could see with infrared light, you would see an image like the one on the right. In this comparison of two images we've got a visible light picture of the constellation Orion on the left and an infrared image of Orion on the right. In it, we see a large, glowing nebula in the sword of Orion. Hidden inside this large, glowing cloud are thousands of newborn stars. In that infrared image, you might've noticed not only the bright glow where the Orion Nebula is, but a bright glow just above it near the star Alnitak, the end star in the belt of Orion.

Stretching about a degree south from Alnitak is a band of nebulosity. In front of this nebulosity is the famous Horsehead Nebula. The Horsehead is a little finger of dusty material that blocks our view of the stars behind. But it's important to know that the Horsehead Nebula is an exceedingly difficult object to observe. It requires a 12-inch telescope or larger under the best of dark skies, and even then a nebular filter may be required to make out the ghostly silhouette of the horse's head.

Let's turn our attention now from Orion to his two hunting dogs. Canis Major looks remarkably like a stick figure drawing of a dog seen in profile. Here are the stars of Canis Major and this is one way that you can make a dog out of the stars. Nearby, Canis Minor, however, is a little bit more than a few stars. There's really just bright Procyon and a couple of adjacent stars that form Canis Minor.

Both dogs are the hunting companions of Orion. Dominating Canis Major is the brilliant star Sirius, also known as the "Dog Star." At magnitude −1.47, it's the brightest star in our night sky. It's an A1 class star at a distance of only 8.6 light years away, making it one of the closest star systems to us, which is one of the reasons it's the brightest star in our sky.

Its name comes from the Greek *seirius*, which means "the scorching one" or "the brilliant one." In Greek times, at the start of the hottest days of summer,

Sirius rose just before the Sun rose in the sky. These hottest days, which are known as the "dog days of summer," were believed to be caused by the additional heat of Sirius, the Dog Star, added to that of the Sun.

Most remarkable of all is not the bright star Sirius A, but its small companion called "Sirius B," or "the Pup." The companion was first seen through a telescope in 1862 and it was surprisingly faint. In this Hubble Space Telescope photograph, the companion is the very faint star to the lower left of Sirius A. This companion is visible in moderate to large telescopes, under conditions when the sky is very, very steady. It's normally difficult to see because of the glare of Sirius A, so you want a nice steady night when Sirius A appears as small as possible.

The companion to Sirius A is a white dwarf star. White dwarf stars are the burned-out cores of low to moderate mass stars. In about six billion years, our Sun will become a white dwarf star. The companion to Sirius has about the same mass as our Sun crammed into a ball about the same size as the Earth, which is 109 times smaller than the Sun. Remember if you took all of the mass in the Sun and crammed it into a ball about the size of the Earth, it would have an incredible density, and in fact white dwarf stars have a density such that a sugar cube of material would weigh about 1.7 tons.

Now that we've seen Canis Major and Canis Minor, which are to the lower left and left of Orion, the east, let's see what's on the other side of Orion.

Let's follow the belt stars up and to the right. If you trace the belt stars of Orion about 20 degrees to the northwest, you'll come to Taurus the Bull. This "V"-shaped grouping of stars is called the "Hyades," and it forms the face of the bull. The brightest star, Aldebaran, forms the eye of the bull, while two stars off on the northeast form the horns of the bull. The Pleiades sit on the shoulder of the bull.

In mythology, Taurus the Bull is another disguise that Jupiter assumes in one of his extramarital affairs. To seduce beautiful Europa, Jupiter changed himself into the likeness of a snow white bull. Europa climbed on the back of the bull, which swam off to Crete where Jupiter ravished Europa. Jupiter gave Europa some gifts, one of which, Laelaps the dog, eventually became

Canis Major. Later, Jupiter placed an image of the bull in the sky where it is today.

In another seduction myth, the bull represents Io. After their affair, Jupiter turned Io into a heifer to hide her from his wife, Juno. You can remember the names of Jupiter's moons because they're all named for his lovers and his favorites: Io, Europa, Ganymede, and Callisto.

The brightest star in Taurus is Aldebaran. Its name means "the follower" in Arabic because Aldebaran follows the Pleiades across the sky. It's a nearby, cool, K3 giant star only 67 light years from the Sun. It has a distinct reddish-orange tint to it.

Taurus is also the home to two nearby clusters of stars. Star clusters come in two types: open clusters and globular clusters. Open clusters, like the one pictured here, typically contain about 50 to 1,000 stars. They're primarily found along the plane of the Milky Way. They're small, usually less than about 30 light years across, and they contain many young, hot, blue stars whose lives you'll remember are very short.

Globular clusters on the other hand are much bigger. They contain between 10,000 and 1 million stars and they're usually found not in the plane of the Milky Way, but surrounding it in all directions in a giant halo, but concentrated down towards the center of our galaxy. Globular clusters are old and they contain no young, hot, blue stars and they're typically 50 to 300 light years across. We know of about 150 globular clusters around our galaxy.

Clusters are an essential place where astronomers study how stars are born, how they live out their lives, and how they die. The great thing about star clusters is that all the stars in a cluster are the same distance away from us. So if a star in a cluster is brighter than the others it means that it's also more luminous. The stars in a cluster were all born at the same time, so they all have the same age. They were all born from the same cloud of gas and dust, so they have the same composition.

You might ask yourself if all the stars in the cluster are the same distance, the same age, and if they're all born from the same material, why are they different? The answer is that stars, when they're born, have different masses. We have small, low mass stars and very large, high mass stars. When a star is born with a given amount of mass, it stays that way for the rest of its life.

The two clusters in Taurus are both open clusters. The first one is the Pleiades, or M45, which appear as a small grouping of six stars just west of the "V"-shaped face of Taurus the Bull. The Pleiades are recognized throughout the world because of their distinctive shape and this tight grouping of moderately bright stars. Most people when they look at the Pleiades see six stars. The mythology surrounding the Pleiades is vast and complex and nearly every society has a story about them.

For example, in Greek mythology the stars represent the Seven Sisters, daughters of Atlas and Pleione. Orion, seeing the seven beautiful sisters, begins pursing them. The sisters cry out to Jupiter to save them, so he changes them into pigeons or doves and they fly up into the sky. Six of the seven sisters consorted with gods, but the remaining sister, Merope, lay with Sisyphus, a mortal. For this reason, her light is dim and she can't be seen. Although the cluster is known as the "Seven Sisters," you can see only six stars.

Another legend tells us that Electra removed herself from the Pleiades because of her grief over the exile of her descendents after the fall of Troy. When Orion died and was placed in the stars, he could again chase after the Pleiades. Every night, as the evening goes on, you will see the Pleiades rise and Orion will follow them into the sky. Then the Pleiades will run across the sky and Orion will forever chase behind them all night long.

The Native Americans, interestingly, also have a myth that the Pleiades represent seven dancing children. One autumn, a band of Onondaga Iroquois made their winter camp. Eight children would sneak away to dance every day instead of doing their chores. They were warned to stop, but they didn't, and one day they floated up into the sky. The son of the chief looked down, and he fell to earth, becoming a shooting star. But the remaining seven floated up to become the Pleiades.

It's remarkable that many cultures, widely separated, believe that there should be seven stars in the Pleiades, but we can only see six. This has led to the idea of a "lost Pleiad," that is, that one of the stars must have been substantially brighter in the past and is dimmer today. Some of the stars in the Pleiades are variable, but none of them vary significantly enough to explain this universal idea of a missing Pleiad. So for now it's still a mystery as to why everyone seems to think there should've been seven sisters or seven children.

The Pleiades are important as a calendar marker throughout the world. In his *Works and Days*, written around 650 B.C., Hesiod tells us that the "Pleiades, daughters of Atlas when they're rising, begin the harvest, and the plowing when they set." The Native America Navajo people called the Pleiades "Dilyehe." When Dilyehe disappears, that is when the Sun moves into this part of the sky and you can no longer see the Pleiades, it's time to start planting the corn. When Dilyehe reappears in the morning sky, it's time to stop planting the corn.

The Pleiades are best viewed with binoculars or with a very low power, wide field telescope. In binoculars, the number of stars in the cluster rises from six to a few dozen. The Pleiades are so large that under moderate or high magnifications you can only see a couple of stars at a time. It's really best seen under low magnification, wide field, or with binoculars. In long exposure photographs, we see hundreds of stars in the cluster.

The cluster is about 435 light years from Earth and most of the bright stars are hot, blue, B-type stars. Within recent times the Pleiades seem to have wandered into a thin cloud of dust. As the dust scatters the blue light of the stars, it creates a faint blue nebulosity that can be seen around the brighter members under dark skies in a low-powered telescope or binoculars. We call this kind of blue glow a "reflection nebula."

Coming back to the "V"-shaped face of the bull we find another nearby open cluster called the "Hyades." In fact, most of the stars in the "V" shape of the face are members of this cluster of stars except for Aldebaran, which happens to be a foreground object. In mythology, the Hyades were the daughters of Atlas and the half-sisters of the Pleiades. When Hyas, their brother, died, they were placed in the sky and turned into stars because of their grief. The

Hyades stretch across 5 degrees of the sky and it's so spread out, that you'll need wide field binoculars in order to appreciate its beauty.

The Hyades are, arguably, one of the most important clusters in the sky. That's because astronomers have used it to calibrate the celestial distance ladder. Astronomers measure the distance to nearby stars using parallax, which is the same way that you measure distances with your two eyes. Imagine viewing your finger out at arm's length. When I view my finger first with my left eye and then with my right eye, and I alternate my eyes back and forth, my finger appears to jump back and forth. Of course, I'm holding my finger still, the reason it appears to jump back and forth is I'm seeing it from two different perspectives.

The change in perspective between my two eyes tells me the distance to my finger. If I bring my finger closer to my eyes and I alternate my eyes, the change in perspective is even larger. As the Earth goes around the Sun, we see the nearby stars from two different perspectives, now when the Earth is on one side of the Sun and then six months later when the Earth is on the opposite side of the Sun. So as we move back and forth around the Sun, the nearby stars appear to shift back and forth, just like my finger was shifting back and forth as I saw it from two different perspectives.

The size of the shift of the position of the star can tell us the distance to the stars. For the nearest stars, this back and forth shift in the sky is about three-quarters of an arcsecond during the year. As the stars get farther away, the shift gets even smaller.

Beyond about 300 light years, the parallax shift in a star is too small to measure accurately. Astronomers have to use the information that we've learned about the nearby stars to determine the distances to those stars. Remember that the brightness of a star in our sky is related to its luminosity, how much energy it gives off every second, and its distance. The closer a star the brighter it is, the further the star the dimmer it is. The more luminous it is the brighter it appears, and the less luminous the dimmer it appears.

These three quantities are related to one another and if we know any two we can determine the third. If we know the luminosity of a star and we can measure its brightness in the sky, we can get its distance. Thanks to these nearby stars whose distances we've been able to measure, we've been able to calculate the luminosities of thousands of nearby stars. We use the luminosities of those nearby stars to get the distances to stars that are further away.

Imagine that there's a star that's so far away that we can't use parallax to measure its distance. We assume that it has the same luminosity giving off the same amount of energy every second as a nearby star of the same type. Once we have its luminosity and measure its brightness, we can calculate the distance to the star.

Beyond these stars, there are other techniques for measuring distances, such as using the period-luminosity relationship for Cepheid variable stars that we discussed in Lecture 7. All these techniques are calibrated using the nearby stars. One of the most important clusters for doing this are the Hyades, and through a variety of methods, astronomers have found that the Hyades are 151 light years away.

Stars live their lives by fusing hydrogen to helium and they have a limited supply of fuel. When they run out of fuel, massive stars explode in a titanic supernova explosion. For a few days around maximum brightness one of these exploding supermassive stars can outshine all the other stars in the galaxy.

Right here in Taurus the Bull we have an example of one of these exploded stars. It's called the "Crab Nebula," M1, which is about 1 degree northwest of Zeta Tauri. It's too faint to see with the eye, but it can be seen in even a small telescope and through a 4- to 6-inch telescope it appears as a faint, fuzzy patch about 4-by-6 arcminutes in size. Through larger telescopes, you can see a few details. The name "M1" tells us it's the first object in Charles Messier's catalog.

Though it's not a spectacular object to look at through a telescope, it's certainly one of the most interesting and that's because M1 is a supernova. Light from this explosion first reached Earth on July 4th in the year 1054. We know this because Chinese astronomers kept meticulous records of the sky and they recorded that on that date a new guest star appeared. It was bright enough that it was seen in full daylight for three weeks and it was visible in the night sky for over a year.

Today when we look at the position recorded by Chinese astronomers we see the exploded remains of the star and they're expanding outward at about 10,000 km/s (about 6,000 miles every second). M1 is about 6,500 light years away so the star didn't explode on July 4th in the year 1054, that's simply when the light reached us. It actually exploded about 6,500 years before that.

Let's turn our attention now to another one of the zodiacal constellations this season. About 40 degrees northeast of Betelgeuse, we come to the pair of stars, Castor and Pollux. They form the heads of twin brothers and it's easy to see two twin brothers in this part of the sky. Imagine two brothers standing next to one another, embracing. Castor and Pollux were the twin brothers of Helen of Troy. The two stars represent the heads of the brothers. A line of stars stretching down from either of the two heads represent the bodies of the brothers, and then faint stars on either side represent their arms.

Their mother, Leda, was visited by Jupiter who disguised himself as a swan. Cygnus the Swan is pictured in our summer sky which is very distant from this part of the sky. That same night, Leda slept with her husband. Two children were born from that evening and Jupiter was the father of Pollux, while her husband was the father of Castor. Pollux was immortal while Castor was mortal. The two brothers grew up together and they were best friends. They were both Argonauts and they sailed with Jason in the quest for the Golden Fleece.

In an argument over women or cattle, depending on the story, Castor was killed. Grief stricken, Pollux asked Jupiter if he could share his immortality with his brother. Jupiter allowed the brothers to spend alternate days on Mt. Olympus and in Hades. Jupiter placed them both in the sky as a symbol of their brotherly love.

Many people confuse the two stars Castor and Pollux. A mnemonic that I use is that "C" comes before "P" in the alphabet, so Castor rises before Pollux.

Another one that you might be able to use is to recall that Castor dies first in the story and Pollux tries to follow him into Hades so Castor will descend down into the horizon first and Pollux will follow. You could also remember that Pollux was the immortal brother and therefore the brighter of the two stars.

Castor and Pollux also have distinct colors. Pollux is a K class, red-orange, giant star while Castor is a blue-white star. Pollux is about 34 light years away and Castor is about 51 light years from the Sun.

Also in the constellation of Gemini the Twins, and visible with the naked eye, is the bright open cluster M35. Under dark skies if you look in the foot of one of the twins you'll see this faint fuzzy patch. With a wide field, low power eyepiece you see a splendid grouping of a few dozen stars. M35 is about 2,800 light years from the Sun and is about 30 light years across. It's a young open cluster. It's about 150 million years old and it contains about 2,500 stars. Its light is dominated by hot, blue stars. The cluster is not old enough yet that all these young, hot, blue stars have died.

While you're looking in Gemini this time of the year remember that the annual Geminid meteor shower is visible around December 13th of every year. The actually date of the peak can shift by a day or two depending on the Earth's orbit around the Sun and how it intersects the stream of particles. Be sure to look on a calendar or in an astronomy magazine or on the Internet to find out when it will be visible this year. The meteors appear to streak out of Gemini and so look in this part of the sky for the show.

Cancer is the constellation that has disappeared in our modern, bright, light-polluted skies. It's often described as being in the empty area between Gemini and Leo. Its brightest star is only magnitude 3.5. In the time of the ancient Greeks, about 2,500 years ago, the Sun was in Cancer on the day of the summer solstice, when it was 23.5 degrees north of the equator. At that latitude the Sun was straight overhead on the summer solstice and so that latitude is known as the "Tropic of Cancer." Today, thanks to the wobbling of the Earth, the sun is actually in Gemini and Taurus on the solstice about 2.5 degrees west of M35.

In Cancer the Crab we find Praesepe, the Beehive Cluster, or M44, and it's visible as a faint naked eye object in the center of Cancer. It's an open cluster about 550 light years away. It's an older cluster about 600 million years old and has about 1,500 stars. The Beehive Cluster is another great, nearby open cluster that you can view in binoculars or a wide field telescope under low powers.

Finally, if we start in Orion and go north we come to Auriga, the Charioteer, who's seen as a large circle of stars north of Taurus. The brightest star in Charioteer is Capella and it's the sixth brightest star in the sky. "Capella" is the Roman name for "she-goat." Next to Capella in the sky is a faint triangle of three stars that are called "the Kids."

In ancient times, the Kids were seen as a separate constellation, but since the time of Ptolemy, they've been part of Auriga. There's no myth explaining why the chariot driver is carrying the goats! Auriga is the home to three very nice open clusters, M36, M37, and M38. Like M35, they're all very nice to view through a telescope at low magnifications.

In the constellations of the winter sky, we've seen the births, lives, and deaths of people, especially Gemini the Twins, and we've also seen the births, lives, and deaths of stars. Though I've titled this lecture "the Winter Sky," you will also see these stars at other times of the year. You'll see them late at night in the fall or just before sunrise in the late summertime.

# The Spring Sky
## Lecture 10

**Again, at this time of the year, I can't tell you where the planets will be when you are out looking at the sky, but we do know that they'll be somewhere along the ecliptic. This time of the year, the ecliptic stretches from the northwest horizon through Gemini, Cancer, Leo, and Virgo. If there's a bright planet out this time of the year, it will be along this line.**

The spring months are a time when the two great celestial bears, Ursa Major and Ursa Minor, are high in the northern sky. Also at this time of the year, we are able to see the bright constellations Leo, Virgo, and Boötes.

The brightest star in Leo is Regulus, but the real treat in this constellation is Algieba, or gamma Leonis, a double star at the nape of the neck of the lion. The radiant of the annual November Leonid meteor shower is near Algieba. Leo also contains two triplets of galaxies. Just to the east of the tail of Leo is the faint constellation Coma Berenices. Here, we find the Coma cluster, which contains more than 1000 galaxies. The north pole of our Milky Way is also located in Coma Berenices.

The constellation Virgo depicts Ceres, whose story in Greek mythology explains why we have seasons when crops flourish and when the soil is barren. Both Spica and Porrima are binary star systems in Virgo. In Porrima, two nearly identical F0 stars are on highly elliptical orbits, which bring them as close together as 5 AUs and as far apart as 81 AUs. Another interesting star in Virgo is Vindemiatrix, the grape gatherer. It rises just before the Sun does, a circumstance known as **heliacal rising**. In August, the heliacal rising of Vindemiatrix marked the time to harvest grapes.

In northern Virgo and southern Coma Berenices, we have an extragalactic wonderland called the Realm of Galaxies. Just the Virgo cluster here is estimated to contain between 1200 and 2000 galaxies. The giant galaxy M87, one of the largest known, is at the core of the Virgo cluster, and at the core of M87 is a supermassive **black hole**, about 6 billion times the mass of our Sun. A black hole is an object whose gravity is so intense that not even light can escape from it. As matter starts to fall into the black hole, it creates an accretion disk surrounding the black hole. From the inner parts of the **accretion disk**, just outside the edge of the black hole, two jets of material are ejected from M87 at speeds close to the speed of light.

**The beauty of the Virgo cluster is the sheer number of galaxies that you can see at one time.**

For most amateur astronomers, the most distant object they have seen through a telescope is the quasar 3C 273, which is located in Virgo. **Quasars** are starlike objects with strong radio emissions. Quasar 3C 273 has a faint magnitude of 12.9, but it is 2 billion light years from the Sun. For an object to appear at that magnitude from a distance of 2 billion light years means it must have a luminosity about 2 trillion times that of our Sun! Today, we believe that 3C 273 is a galaxy that has a supermassive black hole at its center.

Also visible in the spring are the constellation Libra; the bright star Arcturus in the constellation Boötes; and the star Izar, or Pulcherrima, a beautiful double star. Izar offers a wonderful lesson in stellar evolution. How a star lives out its life is determined by its mass. Low-mass stars are cool and red or yellow, while high-mass stars are hot and blue. In this system, the brighter star used to be the more massive of the two, but it has lived out its life and become a yellow star, on its way to becoming a red giant. ■

## Important Terms

**accretion disk**: A rotating disk of gas and dust that forms around the opening of a black hole.

**black hole**: A region of space with gravity so intense that nothing—not even light—can escape from it.

**heliacal rising**: The rising of a celestial body, such as a star, before the Sun.

**quasar**: "Quasi-stellar radio source"; a starlike object with strong radio emissions.

## Suggested Reading

Dickinson, *Nightwatch*, charts 1–5.

Ridpath, *Norton's Star Atlas*, charts 7–10.

Schaaf, *A Year of the Stars*, chaps. 6–8.

## Questions to Consider

1. What is the most distant object you have ever seen with the naked eye? With a telescope?

2. Why is it so difficult to get a good photograph of the whole Virgo cluster?

3. How can we tell that many galaxies harbor supermassive black holes at their centers if black holes emit no light?

4. How many stars or constellations in the sky represent actual, historic people or events?

5. How many of the Messier objects have you seen? Most people are content to slowly work their way through Messier's list of 110 objects, night after night and year after year. Nevertheless, in late March of each year, it is possible to see most, if not all, of the Messier objects in one night. An attempt to see as many as you can in a single night is called a Messier Marathon. Like running a marathon, a Messier Marathon requires practice and training and can be physically challenging. You must be intimately familiar with the locations of all the Messier objects and their surrounding star fields.

# The Spring Sky
## Lecture 10—Transcript

The spring months of March, April, and May are a time when the two great celestial bears, Ursa Major and Ursa Minor, are high in the Northern Sky. Also at this time of the year, the great plain of the Milky Way galaxy is low in the sky. In the Northern Hemisphere, we're looking out of the Milky Way and into deep space and the obscuring clouds of gas and dust in the Milky Way don't block our view of these distant objects.

Thus, this time of the year, the objects we can see through a telescope are dominated by the galaxies and galaxy clusters in the constellations of Virgo and Coma Berenices. We'll also be able to see the bright constellations of Leo, Virgo, and Bootes and we'll be using Ursa Major as our starting point.

To see the spring constellations, set your planisphere for April 15th at 10:00 p.m. The sky will also look like this on March 15th at midnight and May 15th at 8:00 p.m. What you'll see when you set your planisphere is that the familiar constellations of winter which we had discussed in the last lecture are setting over the western horizon. The spring constellations are now high overhead. Orion, Auriga, and Gemini are disappearing and in the center of the window you should see Ursa Major, Leo, Bootes, and Virgo. We'll use the Dipper to find the other constellations in the sky.

From the bowl of the Dipper, take the two stars at the back of the bowl, closest to the handle, and follow them down 40 degrees to come to Leo the Lion, with its bright star Regulus. From Regulus, you can continue the line another 25 degrees to find the faint Watersnake, Hydra, one of Hercules' twelve labors.

If we come back to the Dipper for a minute, follow the arc of the handle down 30 degrees to the bright star Arcturus in Bootes, the Herdsman. You can remember this because you follow the arc of the handle down to Arcturus. If we continue that, take the arc and drive a spike 30 degrees more to the bright star Spica in Virgo, the Virgin. "Spike to Spica" is the way you can remember which star it is. To the lower right of Virgo, that's south and west, we find the small Corvus, the Crow, in the sky. Just to the left or east of Virgo is Libra, the Scales.

Again, at this time of the year I can't tell you where the planets will be when you are out looking at the sky, but we do know that they'll be somewhere along the ecliptic. This time of the year the ecliptic stretches from the northwest horizon, through Gemini, Cancer, Leo, and Virgo. If there's a bright planet out this time of the year, it will be along this line. Now that we've toured the spring sky, let's take a closer look at Leo the Lion.

We draw a line from the two stars at the back of the bowl of the Big Dipper in order to find Leo. These are the two stars that are closest to the handle, Megrez to Phecda and then down another 40 degrees to Regulus, the brightest star in Leo. Leo, I think, looks very much like a lion as seen from the side. The first thing you should find is the large backward question mark with Regulus as the period of the question mark. This question mark is usually known as the "Sickle of Leo the Lion" because of its resemblance to a sickle. The sickle forms the head and the mane of the lion, while Regulus represents the heart of the lion. To the left, or to the east, of the backwards question mark is a triangle of stars that forms the back leg and the tail of Leo.

The first labor of Hercules was to kill the Nemean lion that had been terrorizing Nemea. He was to bring the skin back to Eurystheus, who he had been ordered to serve.

At this time of the year, Hercules is just rising in the east when Leo is high overhead. The lion was the son of the monster Typhon and the brother of the Theban Sphynx, so this was no ordinary lion. In fact, its hide was impervious to weapons. Some say the lion came from the Moon where he'd been suckled by Selene.

Hercules hunted the lion in the forest of Nemea armed with bows and arrows and with a club that he had made from an olive tree that he pulled out of the ground with his bare hands. When the lion attacked, Hercules discovered that his sharp arrows and his weapons had no effect. So he chased the lion to a cave with two entrances; he blocked one and Hercules strangled the lion with his bare hands.

Hercules found that that skin of the lion was so tough that he couldn't use his blades to cut it off. Instead, he was forced to use one of the lion's claws to remove it. Afterwards, Hercules wore the impenetrable skin for protection and Jupiter placed the lion back in the sky from where it came. Now that we know why we have a lion up there in the sky, let's talk about some of the stars that make up Leo the Lion.

The brightest star in Leo is Regulus. It's a B7, main sequence star about 79 light years from our Sun. The real treat, though, in Leo is Algieba, or Gamma Leonis, which is at the nape of the neck of the lion. The radiant of the annual November leonid meteor shower is near Algieba. It's a fine double star, with two components of magnitudes 2.3 and 3.5 separated by 4.4 seconds of arc. The stars have been described as gold-yellow or gold-orange in color. The two stars are about 126 light years away from us and their separation means that they're about 100 astronomical units apart, about twice the distance from the Sun to Pluto.

Leo also contains two triplets of galaxies. All six of the galaxies are about 26 million light years away. You'll need a telescope to see these galaxies, but you should start with a low power eyepiece so that you can fit more than one galaxy in the field at a time. The first triplet are M95, M96, and M105, it's south of the center of Leo. M95 is a spiral galaxy, M96 is a barred spiral galaxy, and M105 is an elliptical galaxy. The second triplet that's visible in Leo are M65, M66, and NGC 3628, all are a few degrees southeast of Theta Leonis and all three of these are spiral galaxies.

Leo Minor is a small constellation just north of Leo. I like to think of it as a little kite in the sky, four stars form a crooked square with a little line coming off that forms the tail of the kite. It's not one of our classical constellations, but it is one of our modern 88. It was introduced by the Polish astronomer Johannes Hevelius in 1687. Its brightest stars are only 4th magnitude so it's a very difficult constellation to see. I bring it up here because of the curious fact that it has no alpha star. Instead, the brightest star in Leo Minor is Beta.

Just to the left, or the east, of the tail of Leo the Lion, which is marked by the star Denebola, lies the faint constellation of Coma Berenices. The constellation contains a large, nearby open cluster of stars that appears as a diffuse fuzzy patch of light with the naked eye. This diffuse patch was variously known as the tuft on the tail of Leo the Lion or the hair of Berenice.

The ancient Greeks didn't count this group of stars as a separate constellation. They thought of it as a part of Leo. But in 1536, it appeared as a separate constellation on a globe of the sky made by the German Caspar Vopel and it was included in Tycho Brahe's influential star catalog of 1602. So we consider it a separate constellation today. Queen Berenice was married to King Ptolemy III Euergetes. Shortly after their marriage, in 246 B.C., Ptolemy headed off to war in Syria to avenge his sister.

Berenice promised the goddess Aphrodite that she would cut off all her beautiful hair and place it on the altar in the temple if her husband were to return safely from war. When he did return, Berenice did as she promised; she cut off her beautiful hair and placed it on the altar.

The next day, the king went by the temple and was angered to find that the hair was missing. He was going to punish the priests of the temple when the court astronomer, Conon of Samos, told Ptolemy that Jupiter had come down from heaven and placed the hair in the sky. Conon pointed to the Coma Cluster of stars and said that that was Berenice's hair glowing like a net in the sky. This faint constellation is the only constellation in the sky whose story concerns a real person.

Berenice's hair is the Coma Star Cluster which consists of a few dozen stars and is about 280 light years from the Sun. Don't confuse this cluster of stars with the nearby Coma Cluster of Galaxies, which is also in Coma Berenice's. The Coma Cluster is a huge cluster of over 1,000 galaxies located about 320 million light years away. The cluster itself is about 20 million light years across and is one of the densest clusters known.

In this image of the Coma Cluster, nearly everything that you see in this image is a galaxy made up of billions of stars. The north pole of our Milky Way is located in Coma Berenices, not far from the Coma Cluster. When we look in this part of the sky we're looking perpendicular to the disk of the Galaxy. There's little or no absorption by dust grains and the star field is thinner up here. There aren't many stars. So we get an excellent view of the universe beyond our Milky Way. It requires a large telescope to see many of the galaxies in the Coma Cluster. In our next constellation, Virgo, we will see a nearby galaxy cluster with dozens of members that are easy to see.

Starting with the Big Dipper, we can follow the arc of the handle down through Arcturus and then spike onto Spica. Spica means an "ear of grain," which is appropriate for a constellation that's associated with Ceres, the goddess of agriculture and grain. I usually find Virgo by looking for the large "Y" that stretches to the upper right, or to the northwest from Spica.

Let me tell you a classical Greek story about Virgo. It explains why we have seasons, which is appropriate for a constellation that's visible in the spring as farmers are planting their crops. Virgo depicts Ceres, who gave birth to Proserpina after an affair with Jupiter. Proserpina was out gathering flowers with her companions when Pluto came along. Pluto had been shot by one of Cupid's arrows and he saw Proserpina, fell in love with her, abducted her, and took her to the underworld, Hades, to spend eternity with him.

Ceres searched the heavens and the earth for her daughter, but she couldn't find her. In her sorrow and her anger, she cursed the ground and the crops stopped growing. In desperation, Ceres asked the Great Celestial Bear, who never sets, if she had seen what happened to her daughter. Remember from Lecture 7 that the celestial bear Callisto is a female hunter and she's high in the sky at this time of the year. The celestial bear replied that the abduction happened in daylight and that she should ask the Sun, who informed her that it was Pluto who abducted her daughter.

Ceres asked Jupiter to secure Proserpina's release from Hades. Jupiter agreed provided that Proserpina had not eaten any food while in Hades. If she had, the fates would forbid her release. Jupiter sent Mercury to retrieve Proserpina but, unfortunately, she had already eaten a few pomegranate seeds and she couldn't rejoin the living. Nevertheless, Jupiter needed to please Ceres in order to make the crops grow again.

He and Pluto reached a compromise. In the spring, Proserpina would leave Hades to join her mother above ground. For the next six months, Ceres is happy to be with her daughter and the crops flourish. Then Proserpina must join her husband in Hades for part of the year. The next months the crops don't grow because of Ceres' sorrow. She doesn't nourish the earth and the crops don't grow.

In Virgo, we have a few bright stars and the one star that we want to talk about in detail is Spica which means an "ear of grain." It's usually drawn as an ear of grain on the classical star maps that Ceres is holding in her hand. The star itself is about 250 light years away and it's a spectroscopic binary of two hot, blue B type stars orbiting one another every few days. A spectroscopic binary means they're too close together for us to see.

Another bright star in Virgo that's interesting is Porrima. Porrima is named for the goddess of prophecy and it's one of the finest double stars in the sky. It's only 39 light years away from us. There are two nearly identical F0 stars in orbit around one another every 169 years. Since they orbit one another every 169 years that means in the course of a few years, it's possible to note a change in the separation and the angle with which the two stars are seen in the sky.

The two stars are on highly elliptical orbits, which brings them as close together as 5 astronomical units and as far apart as 81. They were last closest together in 2005, when they were so close together they were difficult to separate. They were less than half an arcsecond apart. Now is a great time to watch Porrima because over the next few decades, the two stars will continue moving farther apart, not only making them easier to see individually, but making it easy to see the two stars' motions as they orbit one another.

Another interesting star in Virgo is Vindemiatrix. Vindemiatrix's name comes from a corruption of the original Greek and Latin meaning "the grape gatherer." It rises just before the Sun does and we call this the "heliacal rising of a star." In August, at the heliacal rising of Vindemiatrix, the new vintage begins. It marked the time to harvest the grapes.

It's one of the few stars that is mentioned by name in Aratus' poem "Phenomena" and it's by far the faintest of the six stars that he mentioned, the other ones being Arcturus, which is up in the sky right now; Capella, which is just setting in the sky right now though he called it "Aix"; Sirius, which is probably set by this time of the year; Procyon, which may still be in your sky; and Spica right here in Virgo. It's probably mentioned by him because of its importance in determining when to harvest the grapes.

In northern Virgo and southern Coma Berenices we have an extragalactic wonderland called the "Realm of Galaxies." The way to find where all these galaxies are is to draw a giant diamond in the sky, called the "diamond of Virgo." We start with Arcturus in the sky and draw a diamond from Arcturus to Cor Caroli, which is further north in the sky, down to Denebola, which is the tail in Leo the Lion, and down to Spica in Virgo.

On the right-hand side of this diamond, on the border between Virgo and Coma Berenices, it is filled with hundreds of galaxies that are visible in small- and moderate-sized telescopes. Essentially all of these galaxies are part of the Virgo Cluster. This is a plot showing you Denebola, the tail star in Leo the Lion on the right side, and Vindemiatrix over on the left side, the star in Virgo. We can see right between these two stars 100 or more galaxies that you can see in a small telescope.

The Virgo Cluster is estimated to contain between 1,200 and 2,000 galaxies. It's about 60 million light years away from us and it's about 14 million light years across. At the core of the Virgo Cluster are three giant elliptical galaxies, M87, M86, and M84. The beauty of the Virgo Cluster is the sheer number of galaxies that you can see at one time. In a small telescope, you'll see about a half dozen of the brightest members of the cluster. Through an 8-inch telescope, you can see nearly 100. In many fields, more than one galaxy is visible at a time. For example, in this image, the observer saw eight galaxies in one field while centered on M86 and M84. To find your way around this part of the sky you will need a good star atlas to help locate which galaxy is which.

The giant galaxy M87 is at the core of the Virgo Cluster and it's one of the largest galaxies known. It, and the nearby M86 and M84, are giant elliptical galaxies. M87 has a mass of a few trillion times the mass of our Sun. That's ten times the mass of our whole Milky Way galaxy. It's somewhat larger than the Milky Way, about 120,000 light years across. Like all elliptical galaxies,

M87 contains no young, hot, blue stars. Elliptical galaxies went through episodes of star formation long ago. They've used up all of the gas available for young, hot, blue stars. And so elliptical galaxies are primarily composed of old, yellow, orange, and red stars.

At the core of M87, there's a supermassive black hole about six billion times the mass of our Sun. A black hole is an object whose gravity is so intense that not even light can escape from it. As matter falls into the black hole, it starts spinning around the black hole in an effort to fall into the hole and it creates a disk called an "accretion disk" surrounding the black hole. The accretion disk is still outside the black hole, so light can escape from it.

From the inner parts of the accretion disk, just outside the edge of the black hole, two jets of material are ejected from M87 at speeds close to the speed of light. We can see these jets in images of the core of M87. The particles are traveling so fast that they emit tremendous amounts of radio waves, making M87 one of the brightest radio sources in the sky, called "Virgo A," and also it's a bright X-ray source.

In the southern part of Virgo, about 11.5 degrees west of Spica and far from the Virgo Cluster, there's a nearly edge-on spiral galaxy M104, the Sombrero Galaxy. The galaxy may be an outlying member of the Virgo Cluster at a distance from us of about 30 million light years. A small telescope will show that the galaxy is elongated because we're seeing it edge-on. A Hubble Space Telescope image shows us that the disk of the galaxy is cut by a dark dust lane. You can see this dark dust lane in a moderate-sized telescope under dark skies. Here's a picture of what an observer saw. They saw the bright star-like nucleus of the center of the galaxy. Surrounding it is a diffuse glow, but just below the center of the galaxy they could clearly see that the light of the galaxy was cut by this dark lane.

When I give tours of the sky to people, they often ask, what's the most distant object that I've seen with my own eyes through a telescope? The answer for me, as it is for most amateur astronomers, is the quasar 3C 273, which is located right here in Virgo, a few degrees northwest of Gamma Virginis, or Porrima.

The object was first discovered as a bright radio source in the 3rd Cambridge catalog, hence "3C." It's the 3rd Cambridge catalog of bright radio sources in the sky. It was published in 1959 and consisted of a survey of all the bright radio sources. It was the 273rd entry in the catalog so it became known as "3C 273." Once astronomers knew that there was a bright radio source at this location in the sky, they began to hunt it down with optical telescopes to see what kind of object it was.

Given that it was a bright radio source, astronomers were very surprised when they found it optically in the sky and they found that it was just a very faint magnitude 12.9, star-like object. Additional objects like 3C 273 were turned up in other parts of the sky and they became known as quasars, which stands for "Quasi Stellar Radio Sources." They look like stars, but they were incredible sources of radio emission. Most surprising of all is when astronomers broke the light down into its component colors to see what it was made of. They found the familiar lines of hydrogen, but they were shifted tremendously towards the red portion of the spectrum. This meant that the galaxy was not only receding from us very quickly, but that it was also very far away. The object is two billion light years from the Sun.

For an object to appear at magnitude 12.9 from a distance of two billion light years, it must have a luminosity about two trillion times the luminosity of our Sun. That's 200 times the total light output of our whole galaxy! Today we believe that 3C 273 is a galaxy that has a supermassive black hole at its center. Recall that the giant galaxy at the center of the Virgo Cluster, M87, has a supermassive black hole deep down in the center. That black hole was emitting two jets of material in opposite directions as the gas fell into the black hole.

In the case of 3C 273, we believe that the jet is aimed directly at us, so that we're looking right down into the area of the black hole where gas is being swallowed up. As with M87, as the material gathers into a disk, in this accretion disk it gets very, very hot. The gas in that disk is heated to temperatures of millions of degrees. It releases not only visible light, but ultraviolet light and X-rays as well. People often ask me, how can we see light if it's coming from a black hole? Keep in mind that the light we're seeing comes from the gas before it falls into the black hole as it heats up in this disk around the black hole.

Now, 3C 273 is a very difficult object to find because it's magnitude 12.9. You'll need very dark skies and a large telescope 8 to 10 inches or larger. You'll need to print out a detailed finder chart from the Internet, or from one of the sources listed in the resource book that comes with this course. You could also ask a local member of your astronomy club for help. They're likely to have a large telescope and, if they do, they themselves have probably searched out 3C 273 right here in Virgo so that they too can see the most distant object that you can see through even a moderate-sized telescope.

At a distance of two billion light years, we're seeing 3C 273 as it looked two billion years ago. It's so far away that the expansion of the universe is carrying it away from us at 16 percent the speed of light. So not only is this the most distant object you'll ever see, it's also likely the fastest object you'll ever see.

Corvus the Crow is just south and west to the lower right of Spica. It consists of a quadrilateral of four stars. And just west of Corvus is Crater, that's to the right, Crater the Cup, while south of both constellations is the enormous constellation Hydra, which stretches across a huge part of the sky. These three constellations are linked by a classical story.

Each god had a bird that assisted them. For Jupiter it was the eagle, who's represented in the stars of Aquila in the summer sky, and for Apollo, it was the crow. The god Apollo needed water before making a sacrifice to Jupiter so he sent the crow, with a cup, represented by Crater, to get water from a sacred spring. When the crow arrived at the spring, he saw a fig tree whose fruit was not yet ripe. The crow decided to wait a few days for the fruit to ripen and then he ate his fill of fruit.

The crow realized though that this was a mistake because he had made Apollo wait. The crow grabbed a water snake from the spring and told Apollo that the water snake had prevented him from bringing the water sooner. Apollo, who'd been forced to use water from another source, saw right through the lie. He punished the crow by imposing a life of thirst, as reflected in the raspy cry of the crow. To warn us about disobeying the gods Apollo placed the crow and the cup on the back of Hydra in the sky.

As the day goes on, the constellations move from east to west. The Snake and the Cup move across the sky with the thirsty Crow forever chasing behind them. Now Corvus is bright, Hydra and the Cup, Crater, are faint so if you can remember this story once you have found Corvus, just to the lower right of Virgo, you can use it to remember that Crater and Hydra are there as well.

One of the zodiacal constellations that's up this time of the year is Libra. It's one of the dimmer constellations of the zodiac. To find Libra, look between the stars of Virgo and the stars of Scorpius, which are just rising up because they're summer stars. You want to look to the left of Virgo, or off to the east. Libra used to form the front claws of Scorpius, but in Roman times it was split off to become a balance or scales. But in fact, the Romans may have only been reviving an ancient Sumerian constellation because in the Sumerian zodiacal band they had a Balance of Heaven from around 2000 B.C.

Once Libra was turned into a balance, it became associated with the adjacent constellation of Virgo. Virgo is sometimes seen as Dike, the goddess of justice. She lived on Earth during the Golden Age, at a time of peace and happiness when there was no crime or sorrow. When Jupiter overthrew Cronus, the world entered the Silver Age. Jupiter introduced the seasons so that crops wouldn't grow year round. Humans began to argue and they no longer honored the gods.

Dike warned humans that worse to come and she left for the mountains. Next came the Ages of Bronze and Iron, when war and theft and violence were common. Dike left the Earth and joined the constellations where she's pictured as Virgo, next to Libra, judging the decay of mankind.

If you follow the arc of the handle of the Big Dipper down, we follow the arc to down to Arcturus. Arcturus is in the constellation of Bootes, the Herdsman or the Plowman or the Bear Driver. As the name implies Bootes is seen chasing Ursa Major across the sky. If you see the Big Dipper as a plow, you can also think of Bootes as the plowman walking behind the plow.

From Arcturus, you can find a pentagon of stars just north of it and the northernmost star, Beta, forms the head of Bootes. Delta and Gamma form his shoulders, while Epsilon and Rho form his hips. Arcturus is between the legs of Bootes.

There are a number of myths associated with Bootes in the sky. In one of them, Bootes invented the plow that allowed mankind to till the ground better. This pleased Ceres, the goddess of agriculture, who asked Jupiter to place both Bootes and the plow in the sky. In another, Bootes is seen as Arcas, the son of Callisto. We saw in Lecture 7 that Arcas is sometimes pictured as Ursa Minor, but in some accounts, he's Bootes. So in both of these Bootes is associated with Ursa Major either as a bear or the Big Dipper as a plow and Ursa Major is right next to Bootes in the sky.

In a very tragic story, Bootes represents Icarius, a grape grower. Bacchus, the god of wine, was visiting the Earth and viewed the vineyards of Icarius. Bacchus was impressed with what he saw, and taught Icarius how to make wine from grapes. Icarius shared the wine with other shepherds and neighbors who became drunk and fell asleep. When they awoke, they accused him of poisoning them. Icarius' dogs found their master beneath a tree and led his daughter to the body. Distraught, she killed herself as did the dog.

Jupiter placed all of them in the sky as Bootes, his daughter as Virgo, and the dogs as either Canes Major or Canes Minor. Today, we picture the dogs in the nearby constellation of Canes Venatici.

The name of "Arcturus" comes from the Greek for "bear watcher" or "bear guardian." It's the fourth brightest star in our sky and it's 37 light years away from the Sun. Arcturus is unusual in that it's not orbiting the galaxy in the same way as other nearby stars. It lags behind the stars in our part of the galaxy by about 75 miles per second or 120 kilometers per second. It's probably a very old star and either formed early in the history of the galaxy, or it's a captured star that was taken from a cannibalized dwarf galaxy that the Milky Way swallowed up.

The nearby star Epsilon Bootis, Izar, or Pulcherrima is a beautiful double star. Its name "Izar" comes from Arabic for "loin cloth." The Latin "*Pulcherrima*" means it's "most beautiful," and it certainly is. A yellow K star, at magnitude 2.7, it's only 2.8 arcseconds away from a magnitude 5.2 blue, main sequence A star. When you look at the two stars you see a bright yellow K star and a blue B star.

Izar is a wonderful lesson in stellar evolution. How a star lived out its life is determined by its mass. Low mass stars are cool and red and yellow, and high mass stars are hot and blue. In this system, both stars are the same distance away and the brighter star is the yellow star. It's the brighter star because it has already started to die. It used to be the more massive of the two stars. But it has lived out its life and become a red giant or, in this case now, a yellow giant star on its way to becoming a red giant star.

In the spring sky, the Milky Way is low and out of sight for Northern Hemisphere observers. But by the summertime, the Milky Way will be high overhead. As we'll see in the next lecture, the plane of the Milky Way is rich with many sights including star clusters and regions where stars are forming.

# The Summer Sky

## Lecture 11

About 6 billion years from now, our Sun will become a planetary nebula. Its core will shrink down into a white dwarf star, but the atoms in the outer layers of the Sun and the atoms of the inner planets will be spread out amongst the stars, and they'll just wait for that day that gravity pulls them back together again in a star-forming region into the next generation of stars.

The brightest star in the summer sky is Vega in Lyra the Harp, while Deneb is the brightest star in Cygnus the Swan. Stretching south and west from Deneb is the Northern Cross. To the south of Vega is the star Altair in Aquila the Eagle. These three stars, Vega, Altair, and Deneb, form the bright asterism called the Summer Triangle. Also in the summer sky are the constellations Hercules, Corona Borealis, Scorpius, Sagittarius, Ophiuchus the Doctor, and Capricornus.

Infrared observations show that Vega is surrounded by a warm disk of dust, and clumps in the disk suggest that planets may be forming around Vega. The rapid rotation rate of Vega—once every 12.5 hours—has distorted the star so that its equatorial diameter is 23 percent larger than its polar diameter. Near Vega in the sky is epsilon Lyrae, a quadruple star system. Also in Lyra is the Ring Nebula, a planetary nebula, which represents the final stage in stars like our Sun. When such stars die, instead of exploding in a violent **supernova**, they quietly blow off their outer atmospheres in a planetary nebula.

The next bright constellation in the Summer Triangle is Aquila the Eagle, with Altair as its brightest star. A myth from the East relates Altair and Vega as two lovers that were separated by the celestial emperor. Deneb is the third and least bright of the stars that make up the Summer Triangle; it also forms the top of the Northern Cross. The axis of the Northern Cross from Deneb to Albireo forms the body and neck of Cygnus the Swan.

Two other constellations that are high in the summer sky are Corona Borealis (the Northern Crown) and Hercules. As we know, Hercules is the most famous of the Greek heroes, and his 12 Labors have many connections to other constellations. Hercules is the location of the great globular cluster M13, also known as the Hercules cluster. Globular clusters are among the oldest things you will ever see; most are 12 to 13 billion years old, about the same age as the Milky Way.

In the southern part of the sky, we see the distinctive shape of the constellation Scorpius. Within Scorpius is the globular cluster M4, which is much closer to us than most other globular clusters. In addition, Scorpius contains the star 18 Scorpii, a near twin to our Sun in terms of mass, luminosity, age, and rotation period.

Ophiuchus is a huge constellation with no bright stars. In mythology, Ophiuchus is the doctor who cured Orion from the sting of Scorpius. Ophiuchus is also known as the serpent bearer because he splits the constellation Serpens in two. Just east of Scorpius is Sagittarius the Archer, which is most easily found

© Hemera/Thinkstock.

A supernova is a giant explosion at the end of a star's life. Some stars end their lives gently instead, becoming planetary nebulas like the Ring Nebula.

by searching for the Teapot asterism. Sagittarius is one of the richest regions in the sky to observe because it allows us to look down to the center of the Milky Way. After the Orion Nebula, the best star-forming regions visible from the northern sky are all found in Sagittarius. The last of the zodiacal constellations in this part of the sky is Capricornus, the sea goat. ■

## Important Term

**supernova**: An exploding star whose luminosity can increase to many millions of times its normal level.

## Suggested Reading

Dickinson, *Nightwatch*, charts 4, 6–10.

Ridpath, *Norton's Star Atlas*, charts 11–14.

Schaaf, *A Year of the Stars*, chaps. 9–11.

## Questions to Consider

1. What will happen to the Sun when it runs out of fuel about 6 billion years from now?

2. Why is Deneb one of the brightest stars in the sky if it is so far away?

3. In the bright double star Albiero, how do we know that the cool yellow-gold star is more luminous than the hotter blue star? Why is the cooler star the more luminous one?

4. What is the oldest thing you have ever seen?

5. Why are there so many open clusters and star-forming regions visible in the summer sky as compared to the spring sky? Use a planisphere or star map to look at the distribution of star-forming regions.

6. Why are Orion and Scorpius never in the sky at the same time?

# The Summer Sky
## Lecture 11—Transcript

In this lecture we're going to cover the constellations and stars of the summer sky. Because of the nice, warm summer nights, this is the part of the sky that's probably familiar to most people. Arching high overhead we have the most prominent feature in the summer sky, the Milky Way. The deep sky objects this time of year are dominated by the bright star clusters, star forming regions, and dying stars that we find in our galaxy.

We'll begin with a quick tour of the summer sky, and then consider each constellation in turn. We last left off with the spring stars and now that the seasons have moved on the celestial sphere has rotated so that the stars of summer are high overhead. We will start this time with the stars of Ursa Major. So if you have a planisphere, set your planisphere for July 15th at 10:00 p.m. and the stars will look like they will on this map. The sky will also look like this at midnight on June 15, and around 8:00 p.m. in mid August.

First find the Big Dipper setting in the northwestern sky. In particular find the two stars in the bowl that are closest to the handle. You'll remember that these are Gamma and Delta or their proper names are Phecda and Megrez. Draw a line from the star at the bottom of the bowl, Gamma, through Delta, the star at the handle, and continue that line for another 60 degrees. It'll go quite far across the sky over Draco the Dragon and you'll split two bright starts Vega and Deneb.

Vega is the brightest star in Lyra the Harp, and is also the brightest star in our summer sky high over head. Deneb is the brightest star in Cygnus the Swan. Stretching south and west from Deneb is a large cross in the sky called the "Northern Cross." If you drop straight down from Vega to the south, you'll come to the bright star Altair, in Aquila, the Eagle. Altair is somewhat dimmer than Vega, but slightly brighter than Deneb.

These three stars are very important marker stars for the summer sky. Vega, Altair, and Deneb form the bright asterism called the "Summer Triangle." The Summer Triangle is easy to recognize and it's high overhead on summer evenings, and so it's a great way to find other constellations in the sky. To do that we'll use Vega, but we also need to use Arcturus. You'll remember the way to find Arcturus is to follow the arc of the handle of the Big Dipper down to Arcturus.

Now draw a line from Arcturus over to Vega in the Summer Triangle, and you will come across two bright constellations in our summer sky, Hercules and Corona Borealis. Corona Borealis looks like a semi-circle of faint stars, while the center of Hercules looks like a giant trapezoid.

If we go back to the Summer Triangle for a minute, start with Deneb and continue down the length of the Northern Cross, go right through the middle of the Summer Triangle and continue another 80 degrees to the south to the brilliant red-orange star Antares in Scorpius, the scorpion. To the left of Scorpius, to the east, we find the Teapot asterism of Sagittarius, and between Scorpius and Hercules is a huge irregular circle of stars that represents Ophiuchus, the Doctor.

Finally, if you draw a line from Vega to Altair, and continue for another 30 degrees, you'll come to the "V"-shaped constellation Capricornus, the Water Goat, just east of Sagittarius and the last of the zodiacal constellations in the summer. This time of year, the ecliptic is very low in the south and, in fact, if your planisphere plots the southern stars, on the back of the planisphere, you may need to turn it over just to find the ecliptic and it will run through Libra, Scorpius, Sagittarius, and Capricornus.

If there's a bright planet in the summer sky, it will be on the ecliptic. But this time of the year the ecliptic is very low in the south, just above the treetops, which means the planets will be low as well, making them difficult to see down low. If you're using a telescope to look at them, you'll be looking through a lot of atmosphere.

Also, don't forget this time of year the great Perseid meteor shower that peaks around August 12th. The date and the time of the peak varies a little bit from year to year, so be sure to look in an astronomy magazine or online for the best night to go out to see the best of the summer meteor showers. Let's begin with the bright stars and constellations of the Summer Triangle.

Lyra the Harp, or the Lyre, is easy to find thanks to the brilliant star Vega. Next to Vega in the sky is a little parallelogram of four stars and these four stars represent the strings of the harp or the lyre. The lyre is said to have been invented by the god Mercury, who made it from an empty tortoise shell. It was given to Apollo and then to his son, Orpheus, the great mortal musician of his day. Orpheus' wife, Eurydice, died and descended into the underworld. Orpheus decided to get her back by charming Pluto with his lyre.

Pluto agreed to allow Eurydice to follow Orpheus back to the world under the condition that he not look back at her until they had reached the upper air. But near the end of their journey he got worried and he glanced backward, and she was taken away forever. When he died, Jupiter placed the lyre in the sky and Orpheus joined his wife Eurydice in the underworld.

At magnitude 0.0, Vega is the brightest of the summer stars. It's a hot, blue, A0 type star, about 25.3 light years from Earth. Its name is one of the oldest Arabic star names in use and it dates from the 10$^{th}$ century A.D. The name means "the swooping eagle" or "vulture." And rather than a lyre, Arabic astronomers saw a bird in this part of the sky. Infrared observations show us that Vega is surrounded by a warm disk of dust, and clumps in the disk imply that planets may be forming around Vega. It's a very rapidly rotating star. It rotates once every 12.5 hours. To tell you how fast that is, our Sun rotates once every 25 days at the equator. The rapid rotation of Vega has distorted the star so that its equatorial diameter is 23 percent larger than its polar diameter.

Vega has a declination of 38 degrees, 47 minutes. Thus, it passes straight overhead for nearly anyone that lives around 40 degrees north latitude. This is a great marker star to go out in the early evening and see it high in the sky. You'll recall from Lecture 7 that due to precession, this wobble of the Earth, Vega will be a great pole star around 14,000 A.D.

Now near Vega in the sky is Epsilon Lyrae, the "double double." People with excellent vision can see Epsilon Lyrae and that it's a double star. Most people can't do this and they need binoculars or a small telescope to see both components. There are two white stars and their magnitude is 4.6 and 4.7, about 3.5 arcminutes apart. But a moderate telescope at about 100 to 125 times magnifications shows that each of the two stars is actually themselves a double star.

It's rare in the sky that we can see all four stars in a quadruple star system. Usually some of them are so close together we can't make out the individual stars. All four stars in the double double are spectral type A stars, like Vega. They're much fainter than Vega in the sky because now they're much further away, 160 light years away. The stars in each pair are orbiting around one another at a distance of about 140 astronomical units and they take about 1000 years to go around. The pairs of stars are orbiting one another and their orbital period is about 500,000 years or more.

A little more than halfway between Beta and Gamma sits the finest planetary nebula in the Northern Sky. The Ring Nebula, or M57, is small enough that it requires a telescope to see. In fact, at low magnifications or even through your finder scope you might mistake it for a star. But at 150 to 200 times magnification, the Ring Nebula looks like a faint, ghostly green smoke ring in the sky. With bigger telescopes you can even see hints of color in the ring. The Ring Nebula is an example of a planetary nebula, the final stage in stars like our Sun.

When massive stars die, they explode in a violent supernova explosion. But stars like our Sun quietly blow off their outer atmospheres in a spectacular planetary nebula. The name "planetary nebula" is a bit of a misnomer. It comes from the fact that a long time ago when scientists first viewed these through small telescopes, very much like you'll see it in the sky, these nebula look like small, round planets. Of course today, they have nothing at all to do with planets. They're actually the final stage of dying stars.

Just as the story of Lyra reminds us what the Greeks thought would happen after you die, the Ring Nebula reminds us of the fate of the Sun and the inner planets in our solar system. About six billion years from now, our Sun will become a planetary nebula. Its core will shrink down into a white dwarf star. But the atoms in the outer layers of the Sun and the atoms of the inner planets will be spread out amongst the stars and they'll just wait for that day that gravity pulls them back together again in a star-forming region into the next generation of stars.

The next bright constellation in the Summer Triangle is Aquila, the Eagle. Altair is on the northeast side of a diamond that makes up most of the Eagle and Altair is the brightest star in Aquila. Its name comes from the Arabic and it means "the flying eagle" or "vulture" and at 16.8 light years away, Altair is the third closest of the 1st magnitude stars.

In Western mythology, Aquila represents the bird of Jupiter. You might remember from the story of Aquarius that an eagle grabbed the Trojan boy Ganymede and carried him up to Mt. Olympus to be a cup bearer for the gods. Ganymede is depicted in the stars of Aquarius, just to the east of Aquila.

Here's a picture from an old star atlas showing the Aquila, the eagle, carrying a boy off. But in that image the eagle is not carrying Ganymede, it's Antinous. Antinous was a real person, the boy lover of the Roman Emperor Hadrian. Antinous drowned in 130 A.D. and he was commemorated in the stars by Hadrian from the stars southwest of Aquila which previously had not been a constellation. Interestingly, Antinous is no longer one of the constellations in our sky. It's one of a few dozen constellations that were created either by, in this case, a Roman emperor or by astronomers or by cartographers that were added to various star charts in the sky over the years.

When the constellations were settled in the early 20th century by the International Astronomical Union, many of these constellations were left out and we no longer use them in the sky.

My favorite myth from the East is the story relating Altair and Vega. The story is at least 2,500 years old, and it could be much older than that. Altair represents a cowherd while Vega represents the weaving girl or spinning damsel. The two fall in love with one another and they become so distracted that they neglect their celestial duties.

The celestial emperor is displeased and he separates them on opposite banks of a great, impassable celestial river that's represented by the Milky Way. In an image of the summer sky you can see Altair on one side of the galaxy and Vega on the other. Only on the seventh day of the seventh month, all the magpies form a bridge so that the two lovers can cross the Milky Way and be together. But they must return to their respective sides before the dawn comes.

Deneb is the third and least bright of the stars that make up the Summer Triangle. Deneb forms the top of the Northern Cross. It's a large cross of stars that ends in the star Albireo, which sits right near the center of the Summer Triangle. The long axis of the Northern Cross from Deneb to Albireo forms the body and the long neck of Cygnus the Swan, while the shorter perpendicular cross arms from the cross form the outstretched wings of the Swan. It looks very much like a long-necked swan flying to the south in our sky.

The swan in the sky is the great god Jupiter in disguise during one of his many love affairs. In one version of the story, Jupiter desired Nemesis, and when she refused his advances he decided to trick her. He changed into a swan and convinced Venus to change into an eagle and pursue him. He flew to Nemesis who took him in her arms to protect him.

While she was asleep, Jupiter ravished her and flew away, placing the shape of the swan in the sky. From this union, an egg was created from which came beautiful Helen, who later became Helen of Troy. Another version says that Helen had twin brothers, Castor and Pollux, who are now represented in the sky as Gemini, the Twins. Gemini is a winter constellation and it's not visible at this time of the year.

Deneb comes from the Arabic for "tail." As you can see it in the Swan it's in the tail of the Swan. It's an interesting lesson in the bright stars of our night sky. A star can be bright in our sky either because it's close to us or because it's giving off a lot of energy, a lot of light every second. That is, it's a very luminous star.

Sometimes both is true. In the Summer Triangle, for example, we have Vega. It's the fifth brightest star in our sky. It's bright both because it's luminous—it's about 50 times more luminous than our Sun—and it's bright because it's nearby, about 25 light years away. Altair is the same thing. Altair is the 12th brightest star in the sky. It's about ten times the luminosity of our Sun and it's only 17 light years away. Both of these stars are luminous and nearby.

Deneb, on the other hand, is nearly as bright as Altair, but it's 1,425 light years away, nearly 85 times farther away.

When you look at these two stars that appear equally bright, keep in mind that Deneb is 85 times farther. Deneb must be one of the greatest A stars in our Galaxy. It's 54,000 times more luminous than our Sun. If it were placed where Vega is, about 25 light years away, it would shine at magnitude −7.9, 15 times brighter than Venus gets at its brightest. Deneb is a blue, supergiant star. Its radius is about 110 times the diameter of our Sun and in one night Deneb emits as much energy as our Sun will give off in 300 years.

Cygnus is also the home to Albireo, the head of the Swan, which is right at the center of the Summer Triangle. I think it's the most beautiful double star in the sky. The two stars of Albireo are a magnitude 3.1 yellow-gold, K giant star and a magnitude 5.1 , hot, blue, B8 star. The two stars are 34 arcseconds away, relatively far away as binary stars go. They're a spectacular sight in a small telescope.

You'll see the very blue star and the very yellow star right next to one another. And one trick for making the colors of stars a little bit more noticeable is to take the telescope and slightly defocus it a little bit. Sometimes your eyes see the colors a little bit better when the telescope is slightly out of focus.

Now that we've seen the bright Summer Triangle, let's look at the other two constellations that are high up, before we descend to the south to look at the zodiacal constellations this time of year. Remember, to find these two we want to start with Arcturus and draw a line to Vega in the sky. About one-third of the way from Arcturus to Vega, we'll come across Corona Borealis, the circle of stars in the sky. "Corona Borealis" means the "northern crown."

The brightest star in the crown is called Alphecca or Gemma. "Alphecca" comes from the Arabic name which means "to separate" or "broken up," referring to the fact that it's not a complete circle, but it seems to be open on one side. "*Gemma*" is Latin for "gem" or "jewel," which is appropriate since this is the northern crown and it's located right where a jewel would be on the crown.

The circle of stars in the sky represents the fiery gold crown given to Ariadne by her husband Bacchus. In the Native American story that we covered in Lecture 7, you'll remember that Corona Borealis represents the bear's den that he crawls out of at the end of winter.

Given that Hercules is the most famous of the Greek heroes, it's a bit surprising that his constellation isn't brighter and easier to find than it is. To locate Hercules, draw that line from Arcturus to Vega and now go two-thirds of the way from Arcturus to Vega. You'll come to a faint trapezoid of stars called the "Keystone of Hercules." It's because it looks like the keystone that you find at the top of an arch.

The Keystone represents the lower half of Hercules' body. The way you picture Hercules in the sky is to remember that he's standing upside down. His head is pointing to the south and his legs are pointing up to the north. The two stars Delta and Beta represent his shoulders, while the star Alpha, which isn't the brightest in the constellation, is known as Rasalgethi. "Rasalgethi" means the "the kneeler's head." His arms are represented by two lines of stars coming out from his shoulders.

On the other side of the Keystone, Hercules' legs are represented by two lines of stars that come out of the top. Hercules is always depicted as kneeling with at least one leg bent, and in fact, in the *Almagest*, Ptolemy refers to Hercules as "the kneeler."

Hercules is the greatest of the Greek heroes. He was the son of Jupiter and a mortal woman. Jupiter's wife, Juno, surmised that Hercules was the illegitimate son of Jupiter, and she did everything in her power to make his life difficult. Because of her, Hercules had to complete his famous twelve labors.

There are many connections between the twelve labors and other constellations in the sky. Some say, in fact, that the twelve labors are related to the twelve constellations of the Zodiac and that each labor marks the passage of the Sun through one of the zodiacal constellations. I'm not very sure about this because for some of the labors I think it's really difficult to associate it with a specific zodiacal constellation. Here, I'm only going to relate those labors that have an obvious astronomical association.

As you'll remember from Lecture 10, Hercules' first labor was to kill the Nemean Lion, where it's represented in the sky as Leo. When Hercules is high overhead, Leo is just setting in the west. His second labor was to kill the nine-headed monster Hydra. Every time Hercules cut off a head of Hydra, two new ones grew in its place.

To make the task more difficult, Juno sent a giant crab to attack him. Hercules trampled the crab, and with the assistance of Iolaus, he burned the stumps of the severed heads to keep them from growing back. Hydra and Cancer, the Crab, are both low in the western sky when Hercules is about half way up in the eastern sky. At this time of night, 10:00 o'clock when Hercules is high over head, both Hydra and Cancer have set.

In another task, Hercules captures the Cretan Bull which is represented by Taurus in the sky. The two constellations are on opposite parts of the sky, which means that Taurus isn't visible this time of the year. In his next to final task, you'll remember from Lecture 7 that Hercules was to obtain a golden apple from a tree given to Juno as a wedding present. That tree was guarded by a dragon, which is represented by Draco in the sky. Draco is immediately adjacent to Hercules in the sky right at his feet. After his death, Jupiter placed Hercules up in the sky.

In terms of deep sky objects, Hercules is known as the location of the great globular cluster M13, also known as the "Hercules Cluster" and I think it's the best and the brightest of the northern globulars. M13 is about a third of the way from Eta to Zeta, just outside the Keystone. Under really dark skies it's barely visible with the naked eye. With binoculars or a small telescope, it's a faint round fuzzy patch. With a telescope four inches or larger, you can start to see individual stars in the cluster. In telescopes of eight inches or larger, many individual stars are seen in the outer parts of this globular cluster. The center will always remain as a fuzzy, unresolved patch.

It was discovered by Edmund Halley of Halley's Comet fame, in 1714. It contains a few hundred thousand stars. It's about 25,000 light years away from us and over 150 light years across. Hercules has another fine globular cluster, M92, located in the northern part of the constellation.

As you look at a globular cluster through a telescope, keep in mind that it's likely the oldest thing you will ever see in your life. Most globular clusters are 12 to 13 billion years old. The Sun, the Earth, and all the planets in our solar system formed 4.5 billion years ago. So most globular clusters are about 3 times older than the Sun and the Earth.

The Big Bang occurred about 13.7 billion years ago, and the Milky Way formed about 13 billion years ago. So globular clusters are the same age as the Milky Way, and that's because they were some of the first objects to form as our galaxy formed.

Let's move on now to the southern part of the sky where we'll see the zodiacal constellations stretching across probably just above your treetops if you live at mid-northern latitudes. We'll start at Deneb, one of the bright stars in the Summer Triangle, and we'll draw a line down the length of Cygnus, the Swan, the Northern Cross. We'll go through Albireo, the star that forms the head of the Swan at the center of the Triangle, and continue for another 80 degrees to the brilliant orange-red star Antares.

Antares is in Scorpius, the Scorpion, and it's the constellation that I think has the most distinctive shape of all. Three stars form the claws and the head of the scorpion. Then the body of the scorpion comes down to a long tail which hooks around with a stinger on the very end.

Antares forms the heart of the scorpion. Its name comes from the Greek "anti Ares," which means "like Ares" where Ares is the Greek god of war. The Romans knew Ares as Mars, and thus, the name is a reference to the fact that the orange color of Antares looks very much like the orange color of the planet Mars. Antares is a red, supergiant star about 550 light years away. Like Betelgeuse and Orion it's a huge star. It's so large that if we placed it at the center of our solar system it would swallow up the inner planets Mercury, Venus, Earth, and Mars, and a good portion of the asteroid belt.

Also in Scorpius, just west of Antares, is the globular cluster M4. It's invisible to the naked eye because it's so close to Antares, but it does show up in binoculars or a small telescope. In a telescope, M4 is one of the best globular clusters in the sky because it's very close to us. It's only 7,000 light years away, and that's 3 to 5 times closer than most other globulars. Because it's so close to us it's much easier to see individual stars in M4 than in other globular clusters.

Scorpius also contains two magnificent open clusters, M6, the Butterfly Cluster, and M7 called "Ptolemy's Cluster." You can see both of these clusters with the naked eye. M6 is composed of about 50 stars of magnitude 10, and it's a spectacular object in a small telescope. M7 appears as a very bright knot in the Milky Way. It's so spread out that it's actually difficult to see in a telescope. It's probably best viewed with binoculars. Both M6 and M7 were recorded in Ptolemy's catalog of the sky around 150 A.D., and that's why M7 is known as "Ptolemy's Cluster."

One of the least spectacular objects in Scorpius, 18 Scorpii, is worth a look through a telescope or binoculars. It's due south of a star in Ophiuchus. Ophiuchus is the constellation just north of Scorpius. The star's name is "Yed Prior." You can see it on this image here. Barely inside the boundary of the constellation Scorpius is a very faint star, 18 Scorpii.

What's interesting about it is it's a near twin to our Sun. In fact, in basically all ways in terms of mass, luminosity, age, rotation period, it's nearly identical to our Sun. This star is 45.7 light years away. When you look at 18 Scorpii realize that this is what our Sun looks like from a very modest distance of 45.7 light years.

Ophiuchus is a huge constellation with no bright stars in it. It consists of a rough circle of stars about halfway between Hercules and the bright star Antares in Scorpius. If you draw a line from Scorpius up to Hercules, that big circle that fills in the space is Ophiuchus. There's a wonderful story that relates Ophiuchus to Orion and Scorpius in our sky. Orion once bragged that he could kill all the animals on Earth. Gaia sent a giant scorpion to kill Orion. The scorpion stung Orion and he fell to Earth with a mortal wound.

This battle is represented in the sky. Orion and Scorpius are on opposite sides of the sky. As one sets, the other rises and both are never in the sky at the same time. If you set your planisphere so that Orion is setting in the west at say 10:00 p.m. on April 5th, rotate the planisphere just a little bit say to 11:00 p.m. As you move to 11:00 p.m., you'll see Orion setting while the Scorpion rises in the east.

Fortunately for Orion, Ophiuchus the Doctor has a cure for the scorpion's sting and he saves Orion's life. Ophiuchus is pictured in the sky trampling the scorpion underfoot. In fact it's this foot of Ophiuchus that sticks down and crosses the zodiac in the sky, which is why Ophiuchus is now part of our zodiacal constellations. This saving of Orion is represented in the sky as the constellation Ophiuchus sets in the sky Orion rises, restored to life. Ophiuchus is also known as "the serpent bearer" since he splits the constellation Serpens into a leading half, Serpens Caput, and its trailing half, Serpens Cauda.

Just east of Scorpius is the next zodiacal constellation, Sagittarius the Archer. Sagittarius is best found by searching for the Teapot asterism. Three stars form the handle of the Teapot, three form the top of the Teapot, and two more outline the spout of the Teapot. The tea has come to a boil in the pot, or at least the water has come to a boil in the pot, because the bright star clouds of the Milky Way look like steam rising from the spout of the Teapot.

Sagittarius is a centaur in the sky. I think the easiest way to see the centaur is to look for the bow and arrow and the bent arm of the Archer. Three stars form the bow. These stars form his arm pulling back the string on the bow, and then the final star represents the arrow sticking out from the front of the bow.

Sagittarius is one of the richest regions in the sky to observe because here we're looking right down to the center of our Milky Way galaxy. The view is dominated by giant star clouds that have dark dust lanes cutting across them, dozens of objects such as open clusters, globular clusters, and star forming regions. Unfortunately for those of us that live at mid-northern latitudes, it never gets very high in the sky.

After the Orion Nebula, the best star-forming regions visible from the Northern Sky are all found in Sagittarius. M8, the Lagoon Nebula, I think, is second only to the Orion Nebula in the Northern Sky. Through a telescope, it looks like it split in half by a dark lane, which is the Lagoon and how the nebula got its name. It appears bright on opposite sides of the Lagoon. That bright nebula is illuminated by a very young, hot, O5 star. An O5 star is a very, very hot massive star and you can see it as the brightest star in the brightest part of the cloud. The nebula is about 4,000 to 5,000 light years away.

Now just 1.5 degrees north and a little west of M8 is M20, the Trifid Nebula. Like M8, the Trifid is divided by dark lanes. In this case, the lanes divide the object into three parts, which is where it gets the name Trifid. These lanes are dark clouds of dust that are blocking our view of the nebula behind it. M20 is about 5,000 light years away.

Two other nebula appear in Sagittarius, M16 and M17. Although they're fainter than M8 and M20, both are worth viewing. M17 is known as the "Omega Nebula" or the "Swan Nebula" for the shape of the glowing cloud. It looks very much like a swan in the sky. M16 is actually across the border into Serpens Cauda. It's known as the "Eagle Nebula" and it's the home to this famous Hubble Space Telescope image of the Pillars of Creation.

Finally we'll end up on Capricornus, the last of the zodiacal constellation in this part of the sky. To find Capricornus, draw a line from Vega to Altair and continue that line for one more length and you'll be in Capricornus. To see Capricornus in the sky, look for a giant "V." It looks like a large "V" with a smaller, shallower "V" nested inside of it. This constellation represents the Sea Goat with his head in the west and his tail in the east.

In Sagittarius we're looking deep into the center of the Milky Way galaxy. For those of us that live at mid-northern latitudes, it never gets high in the sky, making it difficult to appreciate the beauty of the center of the Milky Way galaxy. In the next lecture, we'll explore the Southern Sky, where the Milky Way arches high overhead in the winter and puts on one of the best displays in the whole sky.

# The Southern Sky and the Milky Way
## Lecture 12

Many people believe that the Southern Cross is the most beautiful sight in the sky. I can't argue that—it really is a beautiful thing to see—but for me, especially as an astronomer who studies the Milky Way, I think that the Milky Way is the most spectacular sight arching across the southern sky.

In this lecture, we will cover a few highlights of the southern sky, and we will focus on the region below 30° declination. The Southern Cross is one of the most impressive star groupings in the sky. The true name of the constellation is Crux. Crux was just barely visible to the Greeks thousands of years ago, but precession has shifted the sky such that Crux is no longer visible from the latitude of Greece. This constellation is made of four bright stars, and it is one of the best in the Southern Hemisphere for seeing the colors of stars. Crux is also useful for finding the **South Celestial Pole**. Just southeast of Crux is the very dark Coalsack Nebula.

Centaurus surrounds Crux on three sides. It contains alpha Centauri, a double star that is the closest star to our Sun. Another star in this system, Proxima Centauri, is the closest one to us other than the Sun, at a distance of 4.22 light years. Centaurus also contains what is probably the greatest globular cluster in our night sky, omega Centauri. This cluster contains about 5 to 10 million stars. Deep in the center of the cluster, the stars are packed so close together that the distance between them is about 0.1 light year. In addition, Centaurus is home to the galaxy NGC 5128, which may actually be a collision between an elliptical and a spiral galaxy.

In the wintertime in the Southern Hemisphere (June, July, and August), the Milky Way is high overhead, as are the constellations Scorpius and Sagittarius. Just south of Scorpius is the classical Greek constellation Ara the Altar. Some constellations in the southern sky are named for recent inventions, such as Telescopium and Microscopium. Others are named for people and animals that the Greeks would not have known about, such as Indus the Native American Indian or Tucana the Toucan.

In the southern spring, the constellation Tucana offers a view of the Small Cloud of Magellan, which is joined in late spring and early summer by the Large Cloud of Magellan. These two companions to our Milky Way are both irregular galaxies, meaning that they have a lot of gas and dust in addition to stars.

**Since we cannot yet see deep inside of stars, we have no idea when a massive star is about to end its life and blow up as a supernova.**

The southern constellation Argo Navis was formerly the largest constellation in the sky but was split into three constellations—Carina, Puppis, and Vela—by the French astronomer Nicolas LaCaille. The best deep sky sight in Carina is the Carinae Nebula, a giant star-forming region about four times larger than the Orion Nebula.

The Milky Way appears as a broad band of light in the southern sky. If we could get outside the Milky Way, we think it would look very much like the spiral galaxy NGC 4414. At the center of the Milky Way is a vast bulge of stars, most of which are old, cool, yellow stars. Surrounding the bulge is the disk of the galaxy, the very flat disk where we live. A classical Greek myth about the Milky Way ties together many of the stories we have heard throughout this course related to the constellations. ■

## Important Term

**South Celestial Pole**: The point directly above the Earth's South Pole projected into space.

## Suggested Reading

Dickinson, *Nightwatch*, chap. 12.

Heifetz and Tirion, *A Walk through the Southern Skies*.

Ridpath, *Norton's Star Atlas*, charts 15–16.

# The Southern Sky and the Milky Way
## Lecture 12—Transcript

You'll remember that the classical Greek constellations that we see in the sky today come from a catalog compiled by Claudius Ptolemaeus in 150 A.D. Because the Southern Sky wasn't visible from Alexandria, Egypt, where Claudius Ptolemaeus was working he didn't catalog any constellations in the far Southern Sky. In the 15$^{th}$ to 17$^{th}$ centuries, in the age of exploration, when astronomers from Europe went to the Southern Hemisphere they began to map the stars and outline new constellations in the Southern Sky for the first time.

This last lecture is a big lecture. We'll be covering a few highlights from the Southern Sky and we'll be focusing on the region below 30 degrees declination. For most of the course we've been covering the Northern Sky, but now we're going to cover this south polar cap. In the constellation in the lecture on the Northern Sky we were covering just that part of the Northern Sky north of 50 degrees declination.

In the seasonal lectures we've already covered some of the constellations that are in the southern skies, such as Sagittarius and Scorpius and Aquarius for example. Now we'll cover some of the highlights in the rest of the Southern Sky.

Recall from that first lecture the Farnese Atlas, the oldest depiction of the classical constellations with its empty region of sky near the South Pole. We'll be studying not only that blank part of the sky, but also classical constellations like Centaurus and Argo Navis, two important classical constellations that precession has carried south and out of view to those of us in the mid-northern latitudes.

We'll start with one of the most impressive star groupings in the entire sky, the Southern Cross. The Southern Cross is an asterism. The true name of the constellation is "Crux." It's the smallest constellation in the sky, and though it's small, it's bright and easy to find, but you have to be careful. There's another cross nearby, which is larger and fainter, called the "False Cross," and the False Cross can fool you.

To be sure that you have the true Southern Cross, find the two brilliant stars Alpha and Beta Centauri. A line drawn from Alpha to Beta Centauri will point right at the true cross in the sky. Remembering that Alpha and Beta Centuari are nearby and that they point at Crux will make sure that you're looking at the right cross in the sky.

Crux was visible to the Greeks thousands of years ago. It would've just barely risen above their southern horizon. They included the stars of Crux in the constellation of Centaurus, which surrounds Crux on three sides. Ptolemy said that the stars were the rear legs of the Centaur. Precession has shifted the sky such that Crux is no longer visible from the latitude of Greece. In fact in the 1600s, Crux was separated from Centaurus and took on its form as a separate constellation as we have it today.

Since it's a modern constellation, there's no mythology associated with it. It does though appear on the national flags of many Southern Hemisphere nations, including Australia, New Zealand, Brazil, Papua New Guinea, and others.

Crux is made of four bright stars and it's one of the best constellations in the Southern Hemisphere for seeing the colors of stars. Alpha and Beta are very hot, blue stars, while Gamma is a cool, orange star. Since the Crux is one of the most-noticeable formation in the Southern Sky, it's worth knowing the names of these stars.

Alpha Crucis, the brightest and the most southerly of the four stars, on the bottom of the cross closest to the South Pole, is known as "Acrux." The name is simply a contraction of "Alpha" and "Crux." It's a modern name, probably created by navigators in the 19th century when they needed a name for the star. And Acrux is a beautiful multiple star system about 325 light years away. There are two blue B stars separated by about 4 arcseconds.

Beta Crucis, or Mimosa, is a modern name and it probably comes from the tropical plant of the same name. Finally Gamma Crucis, or Gacrux, is the northernmost of the stars and it has a name abbreviated from "Gamma" and "Crux."

Not only is Crux beautiful, it's useful as well. Just as we can use Ursa Major to find the north pole in the sky, Crux can be used to find the south pole in the sky. As you'll recall, the north pole in the sky is marked by a moderately-bright star, Polaris. As you can see in this image of the South Celestial Pole there's no bright star near the South Pole. There is a faint magnitude 5.5 star, Sigma Octantis, about a degree away from the pole, but at magnitude 5.5 it's only visible under dark skies.

The most common technique for finding the South Celestial Pole is to draw a line down the length of Crux. Then find Alpha and Beta Centauri and imagine a line drawn between the two stars. These two stars are sometimes called the "Pointer Stars." If you draw a long line that cuts that line in half at a right angle and continue that line down in the sky and have it meet up with the line from the Southern Cross. These two lines meet at a point in the sky very near the South Celestial Pole. Like the North Celestial Pole, a line dropped from the south pole points due south.

At other times of the year, or different times of night, when Crux and Alpha and Beta Centauri aren't visible, another technique that you can use to find the South Celestial Pole is that it forms a roughly equilateral triangle with the two Clouds of Magellan in the sky. When we view Crux under dark skies, we'll see a black spot in the Milky Way just southeast of Crux. This very dark nebula is called the "Coalsack Nebula" for the obvious reason that it's so dark and black.

The Coalsack is a giant cloud of dust about 600 light years away from us and it's absorbing the light from the background stars. We'll hear more about these dark clouds in a little while.

After you use Alpha and Beta Centauri to confirm that you have the right Southern Cross, use it to find the remaining stars in Centaurus because Centaurus surrounds Crux on the east, north, and west sides. You can also find Centaurus by dropping 40 degrees due south from the bright star Spica in Virgo.

The constellation is one of two centaurs in the sky—Sagittarius is the other one—and this centaur represents Chiron who, like all centaurs was half-man and half-horse. The bright stars Alpha and Beta Centauri represent the centaur's front feet, while the stars of Crux are used to represent his back feet. These stars represent the body of the horse, while these stars on the diagram represent his torso.

Here's a story that links Centaurus to Hydra and Hercules in the sky. Hydra is just north of Centaurus while Hercules is far away in the Northern Sky. Chiron was a teacher to many of the great heroes of mythology, like Achilles, Jason, and Asclepius. Chiron was accidentally wounded in the knee by an arrow that Hercules had dipped in the blood of the Hydra, which was a powerful poison. Chiron suffered terribly from this wound but he was immortal, so he couldn't die. To end his suffering, Jupiter allowed Chiron to transfer his immortality to Prometheus and Jupiter placed Chiron in the sky.

Throughout this course, as I've listed the distances to stars both close to us and far away, you may have been wondering, what's the closest star to our Sun? The answer is right here in the Southern Sky, in Centaurus. Alpha Centauri is the brightest star in Centaurus and it's the third-brightest star in the entire sky. It's also known as "Rigil Kentaurus." You'll remember that "*rigil*" is Arabic for "foot," so remember, think of Rigil the star in Orion, and as we've seen, Alpha Centauri represents the foot of the centaur.

At 4.365 light years away, it's the closest stellar system to our Sun. Through a small telescope, Alpha Centauri is seen to be a double star. Alpha Centauri A, the brighter of the two, is a G2 star just like our Sun. The companion is a cooler K1 star and the two stars orbit one another, and their orbital motion is evident even in just a few years of observing.

In the next decade, the stars will be a relatively wide 4 to 5 arcseconds apart. There's a third star in the system, a faint, 11th magnitude star 2 degrees away. This star is very important. It's called "Proxima Centauri," because of all the stars in the sky other than the Sun, it's the closest one to us, at a distance of 4.22 light years. You might ask, if it's the closest star why is 11th magnitude? And the answer is it's a very cool, dim, red star 20,000 times dimmer than our Sun, which is why it's so faint even though it's the closest.

Centaurus contains what is probably the greatest globular cluster in our night sky. In fact, the globular cluster is so bright that it was assigned a Bayer designation and it's known as "Omega Centauri." Omega Centauri contains about 5 to 10 million stars, which is about 10 times more than most globular clusters. Deep down in the center of the cluster, the stars are packed so close together that the distances between them is about 0.1 of a light year.

Compare that to the distance from the Sun to Alpha Centauri, the nearest star system, which is 4.4 light years away. You can see down at the centers of these globular clusters the stars are packed very close together.

Centaurus is also home to the galaxy NGC 5128, called "Centaurus A." It's one of the most peculiar galaxies in the nearby universe. It's visible in binoculars, but a small- to moderate-size telescope will show a dark dust lane cutting right across the center of the galaxy. Except for the dust lane, the galaxy appears to have the shape and the characteristics of an elliptical galaxy.

Elliptical galaxies aren't supposed to have much gas and dust in them, and certainly not a giant dust lane cutting across the middle. It may be that what we're seeing in Centaurus A is a collision between two galaxies, a gas- and dust-poor elliptical and a very gas-rich spiral galaxy, and we're seeing the remains of the merger of these two galaxies.

To the east of Centaurus is the constellation Lupus, the Wolf, and it's usually depicted as an animal that Centaurus has impaled on a long pole or spear. Some Greek sources, such as Eratosthenes, believed that the animal was to be sacrificed on an altar and the constellation Ara, the Altar, is nearby. It's the next one that we'll cover. In these next few slides I'm going to take you on a quick seasonal tour of the southern constellations and for each one of these constellations I'll tell you the season when they're most visible.

In the wintertime in the Southern Hemisphere, this is June, July, and August, the Milky Way is arcing high overhead. Remember the bright zodiacal constellations of Scorpius and Sagittarius, they're high in the sky as well and we've covered them elsewhere. Just south of Scorpius though is the classical

Greek constellation Ara, the Altar, and the story features the origin of the Milky Way.

Saturn, who was one of the Titans, ruled the universe after overthrowing his father, Uranus. Jupiter and his brothers and sisters built an altar and vowed to overthrow Saturn and the Titans. This altar to Jupiter's victory was decorated with stars. The burning incense that was burning on the altar is seen rising up from the altar and it's depicted as the Milky Way in the sky. After they defeated Saturn and the Titans, Jupiter and his brothers Neptune and Pluto divided up the universe. Jupiter was granted the heavens, Neptune, the sea, and Pluto, the underworld.

As we look at the names of the constellations in the Southern Sky, we can see that these are definitely not Greek constellations. Some of them are named for recent inventions. There's a telescope, Telescopium, Microscopium, and Horologium, the Pendulum Clock. Others are named for real people and animals that the Greeks wouldn't have known about, such as Indus the Native American Indian, or Tucana the Toucan.

Most of these constellations have few or no bright stars in them and they don't have any myths or stories associated with them. They were created in the Age of Exploration when astronomers sailing to the Southern Hemisphere first mapped and charted the Southern Sky and filled it with constellations. There are a couple of names in this that stand out. Two Dutch astronomers, Pieter Dirkszoon Keyser and Frederick de Houtman share credit for creating 12 southern constellations that are named after animals or people. They named these, created these constellations, in the late 16th century.

Another astronomer, a French astronomer by the name Nicolas Louis de LaCaille, was working in South Africa and he named 14 new constellations. He named them after scientific instruments and technological devices and he published them in a map of the Southern Sky in 1756.

In the southern spring, the constellation Tucana offers us a view of both the Small Cloud of Magellan and what I think is the second-best globular cluster in the sky. In late spring and early summer, the Small Cloud of Magellan is joined by the Large Cloud of Magellan high in the sky. First let's take a look at the globular cluster 47 Tucanae. Like Omega Centauri, it's bright enough that not only is it visible with the naked eye, it was given a designation that's usually reserved for stars. It too is a very large globular cluster containing a few million stars in a ball only 120 light years across, about 17,000 light years away.

The Small Magellanic Cloud and its neighbor, the Large Magellanic Cloud, which straddles the border of the constellations Dorado, the Goldfish and Mensa, the Table Mountain, are two companion galaxies to our Milky Way. They were brought to the attention of Western scientists by the reports of Antonio Pigafetta who sailed on Magellan's 1519 voyage around the world. At first they were simply known as the "Large" and "Small Clouds," but sometime after 1800 they became known as the "Large" and "Small Clouds of Magellan."

To the naked eye, the Large and Small clouds look like small cloudy patches in the sky, very much like a small patch of the Milky Way. A large pair of binoculars or a telescope will begin to show their stellar nature, while in a moderate telescope you'll be able to start resolving individual stars and other objects in the clouds. The Large Magellanic Cloud is about 170,000 light years away from us while the Small Magellanic Cloud is even further, about 200,000 light years away. Both of them are irregular galaxies, and like most irregular galaxies, they have a lot of gas and dust in addition to stars.

One of the best examples of a giant star-forming region in our local part of the universe is visible in a telescope in the Large Magellanic Cloud and it's called the "Tarantula Nebula" or "30 Doradus." At the center of the Tarantula Nebula is R136, a huge cluster of young, hot, supermassive stars that's seen in this spectacular Hubble Space Telescope photograph. Most of these stars will end their lives in brilliant supernovae explosions. In fact, in the large Megellanic Cloud we have the most important supernova explosion since one that occurred in the Northern Hemisphere in the year 1604.

The one in the Large Magellanic Cloud occurred in February of 1987. Here's an image of the Large Magellanic Cloud in the region near the Tarantula Nebula and it was taken before the star exploded. Can you see in this image the star that's about to explode as a supernova? If not, you're in good company because astronomers had no idea either. Since we cannot yet see deep inside of stars, we have no idea when a massive star is about to end its life and blow up as a supernova. When they do, these stars are so bright that they can outshine the whole galaxy for a few days. In the case of supernova 1987A, it reached magnitude 3 before finally fading away, which means it was a neat and easy naked eye object. Today though, it's too faint to see with the eye even in a large telescope.

The last of the southern constellations that we will discuss are those that made up Argo Navis, which was formerly the largest constellation in the sky. It was split up by the French Astronomer LaCaille into Carina the Keel, Puppis the Poop deck, and Vela the Sails. Argo Navis originally represented the ship that Jason and the Argonauts used in their voyage to retrieve the Golden Fleece, which you'll remember is represented by Aries the Ram in the sky. The ship was crewed by many of the great Greek heroes represented in the sky, including Castor and Pollux, who are up in Gemini the Twins and Hercules who's in the far Northern Sky.

In Argo we have the second brightest star in our sky, Canopus, at a magnitude −0.7. Canopus had originally been Alpha Argo, but is now in Carina and so it's known as Alpha Carina. It's nearly due south of Sirius, but it's too far south to be seen if you live north of latitude 37 degrees. It's a rare class F yellow-white giant star about 310 light years away. Interesting, the name "Canopus" is a mystery. It's a Greek proper name, but it possibly has some Egyptian influence and it was introduced in the 2nd century B.C.

The best deep sky sight in Carina is the Carinae Nebula, sometimes called "Eta Carinae" or "Eta Carinae Nebula" or the "Great Carina Nebula," and it's a giant star-forming region in our Milky Way galaxy about 7,500 light years away from us and it's about four times larger than Orion. Many Southern Hemisphere observers will tell that it's a better sight than the Orion Nebula.

Inside the Carina Nebula is one of the most massive stars in the Milky Way, Eta Carinae. It's 100 to 150 times the mass of our Sun and about four million times more luminous. It's a huge and unstable star and every now and then it can dramatically brighten. In a huge outburst in 1841, it became the second brightest star in the sky, outshining nearby Canopus. During that 1841 outburst, the star threw off tremendous amounts of material that's visible in this Hubble Space Telescope image.

Many people believe that the Southern Cross is the most beautiful sight in the sky. I can't argue that it really is a beautiful thing to see. But for me, especially as an astronomer who studies the Milky Way, I think that the Milky Way is the most spectacular sight arching across the Southern Sky.

The Milky Way appears as a broad band of light. It's faint and you can't see it under light polluted skies, but under dark skies you can trace the galaxy from one horizon all the way to the other horizon. Cutting across the Milky Way, you will see bright star clouds and dark dust lanes. Our Milky Way Galaxy is a vast collection of stars, gas, and dust, and if we could get outside the Milky Way we think it would look very much like this spiral galaxy NGC 4414. Of course we can't get outside the Milky Way so we don't have a picture of our Milky Way from the outside. If we could, though, we think it would be a giant spiral galaxy just like this one made up of giant clouds of gas and dust and hundreds of billions of stars.

Our Milky Way is about 100,000 light years across from side to side and our Sun is about halfway out in the Milky Way, about 26,000 light years from the center of the galaxy. At the center of the Milky Way is a vast bulge of stars, and most of these stars are old, cool stars, so the bulge is yellow in color. Surrounding the bulge is the disk of the galaxy, the very flat disk where we live, which contains not only stars, but clouds of gas and dust.

If we could see the Milky Way outside, edge-on, it would look like this galaxy, NGC 891, which we can see in Andromeda. But because we live in the galaxy we're seeing it edge-on as well, which is why the galaxy takes on the appearance that it does in the sky. By living in it we're seeing it edge-on just from the inside instead from the outside.

We see the central bulge of the Milky Way as a giant bright cloud of stars in Sagittarius and Scorpius. But Sagittarius and Scorpius are always low in the Northern Sky; and much of the bright middle of the Milky Way never even rises above our horizon. But in the Southern Sky, Sagittarius and Scorpius pass high overhead. And not only is the galaxy high overhead, but now you can see the whole bright center of the Milky Way.

To many civilizations, the Milky Way was seen as a roadway or a river in the sky. In the myths of Vega and Altair we saw that the Chinese regarded these two stars as lovers separated by a river, the Milky Way. I want to share with you a second classical Greek legend about the Milky Way because this one ties together many of the other stories that we've heard about when we've been talking about the constellations. It's the story of Phaeton, whose father was the sun god Helios, or Phoebus Apollo, and whose mother was a mortal.

The story, which comes from Ovid, shows a wonderful knowledge of the sky, both about the placement of the constellations in the sky and the motion of the Sun in the sky. Phaeton doesn't have a star or constellation in the sky, but he's the one, he wanted evidence that Helios was his father. He travelled eastward in the direction of sunrise to the palace of his father Helios. The doors to the palace were adorned with images of the heavens and the Earth, including six constellations of the zodiac on one side and six constellation of the zodiac on the other.

Phaeton asked Helios for proof that he was his father. To prove his paternity, Helios offered to grant any wish that Phaeton wanted granted. Phaeton replied that he wanted to ride the chariot of the Sun through the sky for a day. Helios immediately regretted his offer. Though he couldn't withdraw it, he could try to convince his son to pick another wish. He warned Phaeton of the dangers. No one, not even Jupiter, could control the horses like Helios could. There were other dangers as well.

At the beginning of the day, the road is a steep climb and the horses have to use all their strength to climb up. At midday, when the road is highest, it's high overhead, a frightening distance above the Earth. And at the end of the day, the road comes down very steeply and it takes a steady hand to guide the horses and the chariot. I think this is a beautiful description of the celestial sphere over our heads. It rises very steeply, it's high overhead, and then it sets very steeply.

During the trip, the sky is constantly moving carrying the stars and planets with it. The chariot has to be driven fast enough so that the momentum of the sky doesn't overcome his effort to drive in the opposite direction. Phaeton must avoid the horns of Taurus the Bull, the arrows of Sagittarius the Archer, the jaws of Leo the Lion, and the claws of Scorpius the Scorpion, and the claws of Cancer the Crab. But Phaeton persisted, and Helios had no choice but to grant his son's wish.

He gave Phaeton some final advice. He said stay on the zodiac circle, which runs obliquely through the sky, a reference to the fact that the ecliptic cuts across the equator at an angle of 23.5 degrees. Don't go near the southern pole nor too far to the arctic north. Don't ride too close to the Earth or too close to heaven because you don't want to scorch either one. And don't swerve too far to the right, or you will encounter Serpens, which is north of the ecliptic, and don't go too far left or you'll encounter Ara, the Altar, which is to left or the south of the ecliptic.

The goddess, Dawn, opens the doors and the stars vanish. The last to disappear was the morning star, which refers to Venus at its greatest western elongation. The chariot leaps into the sky and out of control. The chariot rode too far north of its normal path, on the ecliptic, close to the Great and Little Bears in the sky who, because of the heat, wanted to cool themselves in the refreshing waters of the Earth by dipping below the horizon. You'll remember that that was forbidden to them.

Draco, the Dragon, who was usually sluggish because he's up in the Northern Sky near the cold, up by the North Pole, was warmed up and became furious. Bootes fled in confusion and was hampered by the Plough, or Ursa Major.

Phaeton gets afraid because the chariot is out of control, and as he heads south he reaches Scorpius and sees the terrible pincers on the scorpion and its curving tail and he panics and he drops the reins of the horses. The horses run wild. First they run too high in the sky and they strike the stars and they gash the sky. Then they run too close to the Earth. Whole cities and nations are reduced to ash, mountaintops burn, rivers dry up, seas are turned to sand, and the ground cracks.

The Moon looks down and is surprised to see her brother's horses below her. Jupiter had to do something and he had no choice. He took a lightning bolt and struck Phaethon. Phaethon tumbled out of the chariot, his hair on fire, and he left a brilliant streak across the sky like a meteor and fell into the river Eridanus, which is pictured in the sky near Orion. Phaeton's friend, Cycnus, cried out in grief. Cycnus turned into a swan and flew up into the sky where he became Cygnus, the Swan. But he was still wary of Jupiter and his lightning bolts, which is supposedly why swans live on lakes instead of land.

Jupiter descended down to the Earth to survey the damage that had been done by the ride of Phaeton, and while he was in the land of Arcadia, he spies beautiful Callisto. You'll remember that her affair with Jupiter gets Callisto, and Callisto's son by Jupiter, turned into the celestial bears, Ursa Major and Ursa Minor. The scorched path of Phaeton's ride is seen as the Milky Way in our sky. Its path really does arch north, through the constellation of Cassiopeia, near Draco and the celestial Bears, far north of the Sun's annual path on the ecliptic.

Of course, the Sun doesn't pass through all the constellations of the zodiac in a single day. It takes a full year for the Sun to pass through the zodiac. In some ways, it's better to think of the chariot ride as relating to the Sun's annual path in the sky with the Milky Way high above the ecliptic in the Northern Sky.

Throughout this course, we have viewed the black sky as the celestial backdrop against which we see the stars and the constellations and all the bright objects, and that's certainly true of the Milky Way. You can see 6,000 stars with the naked eye, many more with a telescope, and hundreds of deep-sky objects. In this course we've learned the bright constellations and how to use them to find our way around the sky.

Where do you go from here? The next step is to continue using them as your starting point for exploring the rest of the sky. There's so much there to see that it can take a lifetime to see it all. You don't have to learn all 88 constellations because it's the bright constellations that we've covered in this course that are key to finding your way and locating the wonderful objects that fill our sky.

The Milky Way also invites us to change our whole way of looking at the sky. That's because with the Milky Way, the bright objects can become the backdrop, while the dark objects stand out. If you look at an image of the Milky Way like this one, cutting down the length of the Milky Way are a number of dark bands, rifts, and clouds. These dark nebula are giant clouds of gas and dust that block our view of the distant stars. One of the more prominent of these is the Great Rift that cuts through the heart of Cygnus the Swan and Aquila, splitting the Milky Way in two in the Northern Sky.

We've spent most of this course discussing the many real and mythological objects and people and animals that various cultures have put up there in the sky. Let me finish the course with two different traditions for looking at the dark spaces, which you can think of as dark constellations of the Milky Way, where the bright stars become the backdrop. The Incas saw a number of animals among the dark stretches of the Milky Way. Along one stretch, they saw a llama with a young llama feeding from it.

A dark cloud, the Coalsack, down near the Southern Cross forms the head of the llama. A long cloud that stretches north is the neck of the llama. An arching dark space above the Milky Way forms the back of the mother while a lower dark cloud is the baby llama feeding. Note how small Crux looks compared to the two llamas in the sky.

A second tradition for looking at the dark spaces of the Milky Way is one seen by Aboriginal Australians. In Ku-ring-gai Chase National Park, north of Sydney, Australia, there's a rock engraving that depicts an emu with its neck stretched out and its legs swept back. The drawing bears a remarkable likeness to an emu that the Indigenous Australians saw in the sky. The head, neck, wings, and legs of the emu are defined by dark rifts in the Milky Way.

While we're looking at the emu and the Milky Way in our sky, let's pan back and pan around the Milky Way and find some familiar constellations. On the left side, we see the bright star Antares and the constellations of Scorpius and Sagittarius familiar to us from our lecture on the summer sky. Between them is the center of the Milky Way. Scanning along to the northeast, we follow the Milky Way to Cygnus the Swan and the Milky Way continues north to the stars of Cassiopeia and then heads south to Orion and Canis Major from our winter sky.

People often ask me how it is that I became an astronomer, and the answer is, I've always wanted to do it. Since I was a little kid, this is what I've wanted to do, to look at the sky and figure out what's out there in the universe. I hope that you've gotten something out of this course as well and that you enjoy looking at the sky as much as I have.

# The 88 Constellations with Selected Bright Stars

O f the roughly 6000 stars in our night sky that can be seen with the naked eye, only a few hundred have proper names. For the remaining stars, there are dozens of other naming conventions in use, but the most common is to use their Bayer designation. In his star atlas *Uranometria*, published in 1603, Johann Bayer began the practice of labeling the stars in each constellation with a lowercase Greek letter followed by the genitive case of the constellation name. In most cases, the letters were assigned in order of brightness: α was the brightest star in the constellation, β was he second, and so forth. Nevertheless, Bayer did not always follow this convention. In Ursa Major, for example, he listed the stars in order along the asterism we know as the Big Dipper, starting with the end star in the bowl.

The letters of the Greek alphabet appear immediately below. Following the table is a list of the 88 official constellations adopted by the International Astronomical Union in 1922, including the name of the constellation, the genitive form of its name, its abbreviation, its name in English translation, and the names of any bright stars covered in this course in order of their Bayer designation. All but one of the 48 classical constellations compiled by Ptolemy around A.D. 150 are listed and are marked with asterisks; Ptolemy's Argo Navis was divided into the modern constellations Carina, Puppis, and Vela. ■

## Greek alphabet

| | | | | |
|---|---|---|---|---|
| α alpha | ζ zeta | λ lambda | π pi | φ phi |
| β beta | η eta | μ mu | ρ rho | χ chi |
| γ gamma | θ theta | ν nu | σ sigma | ψ psi |
| δ delta | ι iota | ξ xi | τ tau | ω omega |
| ε epsilon | κ kappa | ο omicron | υ upsilon | |

# The 88 Constellations with Selected Bright Stars

**Table 1. The 88 constellations with selected bright stars**

| Constellation | Genetive | Abbreviation | English Name | Major Stars |
|---|---|---|---|---|
| Andromeda* | Andromedae | And | Andromeda | Alpheratz (formerly shared with Pegasus), Mirach, Almach |
| Antlia | Antliae | Ant | Air Pump | |
| Apus | Apodis | Aps | Bird of Paradise | |
| Aquarius* | Aquarii | Aqr | Water Carrier | |
| Aquila* | Aquilae | Aql | Eagle | Altair (also part of Summer Triangle) |
| Ara* | Arae | Ara | Altar | |
| Aries* | Arietis | Ari | Ram | Hamal |
| Auriga* | Aurigae | Aur | Charioteer | Capella |
| Boötes* | Boötis | Boo | Herdsman | Arcturus |
| Caelum | Caeli | Cae | Chisel | |
| Camelopardalis | Camelopardalis | Cam | Giraffe | |
| Cancer* | Cancri | Cnc | Crab | |
| Canes Venatici | Canun Venaticorum | CVn | Hunting Dogs | Cor Caroli |
| Canis Major* | Canis Majoris | CMa | Great Dog | Sirius (brightest star in the sky) |
| Canis Minor* | Canis Minoris | CMi | Little Dog | Procyon |
| Capricornus* | Capricorni | Cap | Sea Goat | |
| Carina | Carinae | Car | Keel | Canopus |

Table 1. The 88 constellations with selected bright stars (cont.)

| Constellation | Genetive | Abbreviation | English Name | Major Stars |
|---|---|---|---|---|
| Cassiopeia* | Cassiopeiae | Cas | Cassiopeia | Shedar, Caph |
| Centaurus* | Centauri | Cen | Centaur | α Centauri (a.k.a. Rigel Kentaurus), Hadar |
| Cepheus* | Cephei | Cep | Cepheus | |
| Cetus* | Ceti | Cet | Sea Monster or Whale | |
| Chamaeleon | Chamaeleontis | Cha | Chameleon | |
| Circinus | Circini | Cir | Compasses | |
| Columba | Columbae | Col | Dove | |
| Coma Berenices | Comae Berenices | Com | Berenice's Hair | |
| Corona Australis* | Coronae Australis | CrA | Southern Crown | |
| Corona Borealis* | Coronae Borealis | CrB | Northern Crown | Alphecca/Gemma |
| Corvus* | Corvi | Crv | Crow | |
| Crater* | Crateris | Crt | Cup | |
| Crux | Crucis | Cru | Southern Cross | Acrux, Mimosa, Gacrux |
| Cygnus* | Cygni | Cyg | Swan | Deneb, Albireo |
| Delphinus* | Delphini | Del | Dolphin | |
| Dorado | Doradus | Dor | Goldfish | |
| Draco* | Draconis | Dra | Dragon | Thuban |
| Equuleus* | Equulei | Equ | Little Horse | |

211

# The 88 Constellations with Selected Bright Stars

## Table 1. The 88 constellations with selected bright stars (cont.)

| Constellation | Genetive | Abbreviation | English Name | Major Stars |
|---|---|---|---|---|
| Eridanus* | Eridani | Eri | River | Achernar |
| Fornax | Fornacis | For | Furnace | |
| Gemini* | Geminorum | Gem | Twins | Castor, Pollux |
| Grus | Gruis | Gru | Crane | |
| Hercules* | Herculis | Her | Hercules | Rasalgethi |
| Horologium | Horologii | Hor | Pendulum Clock | |
| Hydra* | Hydrae | Hya | Water Snake | Alphard |
| Hydrus | Hydri | Hyi | Little Water Snake | |
| Indus | Indi | Ind | Indian | |
| Lacerta | Lacertae | Lac | Lizard | |
| Leo* | Leonis | Leo | Lion | Regulus, Denebola, Algeiba, Zosma |
| Leo Minor | Leonis Minoris | LMi | Little Lion | |
| Lepus* | Leporis | Lep | Hare | |
| Libra* | Librae | Lib | Scales | Zubenelgenubi |
| Lupus* | Lupi | Lup | Wolf | |
| Lynx | Lyncis | Lyn | Lynx | |
| Lyra* | Lyrae | Lyr | Lyre or Harp | Vega |
| Mensa | Mensae | Men | Table Mountain | |
| Microscopium | Microscopii | Mic | Microscope | |

Table 1. The 88 constellations with selected bright stars (cont.)

| Constellation | Genetive | Abbreviation | English Name | Major Stars |
|---|---|---|---|---|
| Monoceros | Monocerotis | Mon | Unicorn | |
| Musca | Muscae | Mus | Fly | |
| Norma | Normae | Nor | Set Square | |
| Octans | Octantis | Oct | Octant | |
| Ophiuchus* | Ophiuchi | Oph | Serpent Bearer | Rasalhague |
| Orion* | Orionis | Ori | Hunter | Betelgeuse, Rigel, Bellatrix, Mintaka, Alnilam, Alnitak, Saiph |
| Pavo | Pavonis | Pav | Peacock | |
| Pegasus* | Pegasi | Peg | Winged Horse | Markab, Scheat, Algenib, Enif |
| Perseus* | Persei | Per | Perseus | Mirfak, Algol |
| Phoenix | Phoenicis | Phe | Phoenix | |
| Pictor | Pictoris | Pic | Painter's Easel | |
| Pisces* | Piscium | Psc | Fishes | |
| Piscis Austrinus* | Piscis Austrini | PsA | Southern Fish | Fomalhaut |
| Puppis | Puppis | Pup | Stern | |
| Pyxis | Pyxidis | Pyx | Compass | |
| Reticulum | Reticuli | Ret | Reticle | |
| Sagitta* | Sagittae | Sge | Arrow | |
| Sagittarius* | Sagittarii | Sgr | Archer | |

# The 88 Constellations with Selected Bright Stars

**Table 1. The 88 constellations with selected bright stars (cont.)**

| Constellation | Abbreviation | Genetive | English Name | Major Stars |
|---|---|---|---|---|
| Scorpius* | Sco | Scorpii | Scorpion | Antares |
| Sculptor | Scl | Sculptoris | Sculptor | |
| Scutum | Sct | Scuti | Shield | |
| Serpens* | Ser | Serpentis | Serpent | |
| Sextans | Sex | Sextantis | Sextant | |
| Taurus* | Tau | Tauri | Bull | Aldebaran, Elnath |
| Telescopium | Tel | Telescopii | Telescope | |
| Triangulum* | Tri | Trianguli | Triangle | |
| Triangulum Australe | TrA | Trianguli Australis | Southern Triangle | |
| Tucana | Tuc | Tucanae | Toucan | |
| Ursa Major* | UMa | Ursae Majoris | Great Bear | Dubhe, Merak, Phecda, Megrez, Alioth, Mizar and Alcor, Alkaid (a.k.a. the stars of the Big Dipper asterism) |
| Ursa Minor* | UMi | Ursae Minoris | Little Bear | Polaris. (The stars of Ursa Minor also form the Little Dipper asterism.) |
| Vela | Vel | Velorum | Sails | |
| Virgo* | Vir | Virginis | Virgin | Spica, Porrima, Vindemiatrix |

**Table 1. The 88 constellations with selected bright stars (cont.)**

| Constellation | Genetive | Abbreviation | English Name | Major Stars |
|---|---|---|---|---|
| Volans | Volantis | Vol | Flying Fish | |
| Vulpecula | Vulpeculae | Vul | Fox | |

*Note:* * = One of Ptolemy's 48 classical constellations.

# The 30 Brightest Stars

The following table lists the 30 brightest stars in the sky arranged in order of apparent brightness. No list of the brightest stars can be definitive since some stars vary in brightness (e.g., Betelgeuse and Antares) and the magnitudes come from a wide variety of sources. Thus, there will always be minor differences from one listing to the next. In the table following, two magnitudes are listed for double stars where each component can be seen in a telescope. For double stars that are too close to be resolved individually, I give the combined magnitude and list the spectral types of each star.

In addition to the proper name of the star and the constellation in which the star is located, the table lists the right ascension (RA), declination (DEC), spectral class, luminosity class, distance, and apparent magnitude. In the case of multiple stars, two magnitudes are given if the individual stars can be separately seen in a telescope. If the two stars are too close together to be separated, a single combined magnitude is given and the spectral types of both stars are listed.

The declination, given in degrees, minutes, and seconds of arc, can be used to determine if the star is ever visible at your latitude. For observers in the northern hemisphere, the most southern declination that can be seen is your latitude minus 90°. For example, at a latitude of 40° north (Madrid, Philadelphia, Denver, Beijing), the most southern declination that is visible at all is 40° − 90° = −50°. This declination just touches the southern horizon. To be easily seen, a star will need to be 5–10° higher. From a latitude of 40° north, Canopus, with a declination of −52.7°, is never visible since it is farther south than −50°. In the southern hemisphere, the most northern declination that is visible is your latitude plus 90°.

The right ascension, given in hours, minutes, and seconds of time, can be used to determine when a star is visible. For example, high in the sky around 10 pm are right ascensions of 3–9$^h$ in the winter (December, January, February), 9–15$^h$ in the spring (March, April May), 15–21$^h$ in the summer

216

(June, July, August), and $21–3^h$ in the fall (September, October, November).

The spectral class of the star is a measure of its temperature. In order from hottest to coolest, the spectra sequence is O, B, A, F, G, K, and M. Each spectral class is subdivided into to subclasses 0 (hotter) through 9 (cooler). Thus, an A8 star is hotter than an A9, which is hotter than an F0.

The luminosity class tells us the size of the star. Luminosity class is usually denoted by a Roman numeral after the spectral type. In order of luminosity from least to greatest, they are the main sequence stars (V), subgiants (IV), giants (III), bright giants (II), and supergiants (I)—which have been subdivided into types luminous supergiants (Ia) and less luminous supergiants (Ib). Main sequence stars are stars that are fusing hydrogen into helium. Since this is, by far, the longest-lasting phase in the life of a star, most stars are on the main sequence.

Although giants and supergiants make up only a small fraction of all stars, they are extraordinarily luminous and comprise the majority of the brightest stars in our sky. As their names imply, giants (III), bright giants (II), and supergiants (Ib and Ia) have large diameters compared to main sequence stars of the same type. They are evolved stars that have exhausted the hydrogen in their cores and are dying. Depending on their evolutionary stage, giant stars can have a shell of hydrogen fusing into helium around an inert helium core undergoing no nuclear reactions; a hydrogen-fusing shell around a core fusing helium to carbon; or a hydrogen-fusing shell around a helium-fusing shell around a core fusing carbon to oxygen.

The bright stars in the right ascension range 3–9 hours are part of the Gould Belt, a true ring of stars whose plane is tilted with respect to the Milky Way by about 18°. The origin of the Gould Belt, first noticed in the mid-19th century, is still debated, but it likely represents the plane of the local star-forming regions in our part of the galaxy.

## Suggested Reading

Kaler, *Stars and Their Spectra.*

# The 30 Brightest Stars

## Table 2. The 30 brightest stars

| Rank | Apparent Magnitude | Proper Name | Constellation | RA h m | | DEC ° m | | Spectral and Luminosity Class | Type of Star | Distance (LY) |
|---|---|---|---|---|---|---|---|---|---|---|
| 1 | −1.46 | Sirius | Canis Major | 06 | 45 | −16 | 42 | A1 V | Main sequence | 8.6 |
| 2 | −0.72 | Canopus | Carina | 06 | 24 | −52 | 41 | F0 Ib–II | Bright giant or supergiant | 309 |
| 3 | −0.01, 1.33 | α Centauri | Centaurus | 14 | 40 | −60 | 50 | G2 V, K1 V | Both main sequence | 4.36 |
| 4 | −0.04 | Arcturus | Boötes | 14 | 16 | +19 | 11 | K1.5 III | Giant | 37 |
| 5 | 0.03 | Vega | Lyra | 18 | 37 | +38 | 47 | A0 V | Main sequence | 25 |
| 6 | 0.08 | Capella | Auriga | 05 | 17 | +46 | 00 | G8 III, G0 III | Both giants | 43 |
| 7 | 0.12 | Rigel | Orion | 05 | 15 | −08 | 12 | B8 Iab | Supergiant | 860 |
| 8 | 0.34 | Procyon | Canis Minor | 07 | 39 | +05 | 14 | F5 IV–V | Subgiant or main sequence | 11.5 |
| 9 | 0.42 | Betelgeuse | Orion | 05 | 55 | +07 | 24 | M2 Iab | Supergiant | 570 |
| 10 | 0.46 | Achernar | Eridanus | 01 | 38 | −57 | 15 | B3 V | Main sequence | 140 |
| 11 | 0.61 | Hadar | Centaurus | 14 | 04 | −60 | 22 | B1 III | Giant | 392 |
| 12 | 0.77 | Altair | Aquila | 19 | 51 | +08 | 52 | A7 IV–V | Subgiant or main sequence | 16.7 |
| 13 | 1.3, 1.8 | Acrux | Crux | 12 | 27 | −63 | 06 | B0.5 IV, B1 V | Subgiant and main sequence | 325 |
| 14 | 0.85 | Aldebaran | Taurus | 04 | 36 | +16 | 30 | K5 III | Giant | 67 |
| 15 | 0.96 | Antares | Scorpius | 16 | 29 | −26 | 26 | M1.5 Ib | Supergiant | 550 |
| 16 | 1.04 | Spica | Virgo | 13 | 25 | −11 | 09 | B1 V, B4 V | Both main sequence | 250 |
| 17 | 1.14 | Pollux | Gemini | 07 | 45 | +28 | 01 | K0 III | Giant | 34 |
| 18 | 1.16 | Fomalhaut | Piscis Austrinus | 22 | 58 | −29 | 37 | A3 V | Main sequence | 25 |
| 19 | 1.25 | Deneb | Cygnus | 20 | 41 | +45 | 16 | A2 Ia | Supergiant | 1425 |

**Table 2. The 30 brightest stars (cont.)**

| Rank | Apparent Magnitude | Proper Name | Constellation | RA h | RA m | DEC ° | DEC m | Spectral and Luminosity Class | Type of Star | Distance (LY) |
|---|---|---|---|---|---|---|---|---|---|---|
| 20 | 1.29 | Mimosa | Crux | 12 | 48 | −59 | 40 | B0.5 III–IV | Giant or subgiant | 280 |
| 21 | 1.35 | Regulus | Leo | 10 | 08 | +11 | 58 | B7 V | Main sequence | 79 |
| 22 | 1.50 | Adhara | Canis Major | 06 | 59 | −28 | 58 | B2 II | Bright giant | 405 |
| 23 | 1.96, 2.91 | Castor | Gemini | 07 | 35 | +31 | 53 | A1 V, A5 V | Both main sequence | 51 |
| 24 | 1.63 | Gacrux | Crux | 12 | 31 | −57 | 07 | M3.5 III | Giant | 88 |
| 25 | 1.63 | Shaula | Scorpius | 17 | 34 | −37 | 06 | B1.5 IV, B2 | Sub-giants | 365 |
| 26 | 1.64 | Bellatrix | Orion | 05 | 25 | +06 | 21 | B2 III | Giants | 245 |
| 27 | 1.65 | Elnath | Taurus | 05 | 26 | +28 | 36 | B7 III | Giants | 130 |
| 28 | 1.68 | Miaplacidus | Carina | 09 | 13 | −69 | 43 | A2 IV | Sub-giants | 111 |
| 29 | 1.70 | Alnilam | Orion | 05 | 36 | −01 | 12 | B0 Ia | Supergiants | 1340 |
| 30 | 1.74, 4.21 | Alnitak | Orion | 05 | 41 | −01 | 57 | O9.5 Ibe, B0 III | Supergiant and giant | 815 |

*Note:* h = hours; m = minutes; ° = degrees; LY = light-years

# Data for the Planets and Dwarf Planets in Our Solar System

## Basic Physical and Orbital Data

Table 3. Basic physical and orbital data for planets and dwarf planets

| Names | Radius (Earth = 1) | Mass (Earth = 1) | Distance from Sun (AU) | Orbital Period (Years) | Rotation Period (Earth Days) |
|---|---|---|---|---|---|
| Mercury | 0.382 | 0.055 | 0.387 | 0.2409 | 58.6 |
| Venus | 0.949 | 0.815 | 0.723 | 0.6152 | −243.0 |
| Earth | 1.00 | 1.00 | 1.00 | 1.00 | 0.9973 |
| Mars | 0.533 | 0.107 | 1.524 | 1.881 | 1.026 |
| Ceres | 0.076 | 0.00016 | 2.77 | 4.60 | 0.378 |
| Jupiter | 11.19 | 317.9 | 5.203 | 11.86 | 0.41 |
| Saturn | 9.46 | 95.18 | 9.539 | 29.46 | 0.44 |
| Uranus | 3.98 | 14.54 | 19.19 | 84.01 | −0.72 |
| Neptune | 3.81 | 17.13 | 30.06 | 164.8 | 0.67 |
| Pluto | 0.181 | 0.0022 | 39.48 | 248.0 | −6.39 |
| Eris | 0.22 | 0.0028 | 67.67 | 557 | 15.8 |

*Note*: Each equatorial radius is given in terms of the Earth's equatorial radius (6378 km), and the mass is given in terms of the Earth's mass ($5.97 \times 10^{24}$ kg). The average distance of the planet from the Sun is given in astronomical units, where 1 AU = $1.496 \times 10^8$ km.

## Visibility Data for the Planets

### Mercury

Mercury is only visible at its greatest elongations from the Sun. Greatest eastern elongations occur in the evening sky when Mercury is visible in the west after sunset. Greatest western elongations occur in the morning sky when Mercury is visible in the east just before sunrise. Since Mercury completes an orbit around the Sun in just under 88 days, there are as many as 7 elongations in a year. The dates of greatest elongations can be found on the Internet or in astronomy magazines. In general, for observers in either hemisphere, greatest eastern (evening) elongations are best in the spring while greatest western elongations (morning) are best in the fall. Mercury is only visible for a few days before and after greatest elongation.

**Venus**

Like Mercury, Venus is most visible during its greatest elongations. Since Venus can appear much farther from the Sun than Mercury (47° at its best) and orbits more slowly, it is easily visible for months around the time of greatest elongation. After an eastern (evening) elongation, the next western (morning) elongation occurs about 20 weeks later. From western elongation back to eastern elongation, the interval is 63 weeks. Thus, Venus can have 0, 1, or 2 elongations in a year. The table below lists the greatest eastern and western elongations of Venus from 2010 to 2031.

Table 4. Elongations of Venus, 2010–2031

| Elongation | Date | Elongation | Date |
|---|---|---|---|
| Eastern | 2010 Aug 20 | Eastern | 2021 Oct 29 |
| Western | 2011 Jan 8 | Western | 2022 Mar 20 |
| Eastern | 2012 Mar 27 | Eastern | 2023 Jun 4 |
| Western | 2012 Aug 15 | Western | 2023 Oct 23 |
| Eastern | 2013 Nov 1 | Eastern | 2025 Jan 10 |
| Western | 2014 Mar 22 | Western | 2025 Jun 1 |
| Eastern | 2015 Jun 6 | Eastern | 2026 Aug 15 |
| Western | 2015 Oct 26 | Western | 2027 Jan 3 |
| Eastern | 2017 Jan 12 | Eastern | 2028 Mar 22 |
| Western | 2017 Jun 3 | Western | 2028 Aug 10 |
| Eastern | 2018 Aug 17 | Eastern | 2029 Oct 27 |
| Western | 2019 Jan 6 | Western | 2030 Mar 17 |
| Eastern | 2020 Mar 24 | Eastern | 2031 Jun 2 |
| Western | 2020 Aug 13 | Western | 2031 Oct 21 |

*Note*: All eastern elongations occur in the evening, and all western elongations occur in the morning.

## Mars

Mars is best visible at opposition, when it is opposite the Sun in the sky (when the Earth passes between Mars and the Sun) and thus highest in the sky at midnight. At opposition, Mars is closet to Earth, so it appears largest and brightest in our sky. Oppositions of Mars occur every 780 days on average. Due to Mars's elliptical orbit, an opposition can occur when Mars is closer or farther from the Sun than average. Perihelic oppositions, which occur when Mars is closest to the Sun, are the best, but all of the perihelic oppositions occur at far southern declinations, when Mars is hard to see from the Northern Hemisphere. Though Mars is visible for a few months around opposition, it is best seen through a telescope within a month of opposition. The table below (from http://www.seds.org) lists the oppositions of Mars between 2010 and 2031.

Table 5. Oppositions of Mars, 2010–2031

| Opposition Date | Minimum Distance from Earth AU | (millions of km) | Maximum Angular Size of Mars (arc seconds) |
|---|---|---|---|
| 2010 Jan 29 | 0.664 | 99.33 | 14.10 |
| 2012 Mar 3 | 0.674 | 100.78 | 13.89 |
| 2014 Apr 8 | 0.618 | 92.39 | 15.16 |
| 2016 May 22 | 0.503 | 75.28 | 18.60 |
| 2018 Jul 27 | 0.385 | 57.59 | 24.31 |
| 2020 Oct 13 | 0.415 | 62.07 | 22.56 |
| 2022 Dec 8 | 0.544 | 81.45 | 17.19 |
| 2025 Jan 16 | 0.642 | 96.08 | 14.57 |
| 2027 Feb 19 | 0.678 | 101.42 | 13.81 |
| 2029 Mar 25 | 0.647 | 96.82 | 14.46 |
| 2031 May 4 | 0.553 | 82.78 | 16.91 |

## Jupiter and Saturn

Jupiter comes to opposition every 399 days. Saturn comes to opposition every 378 days. Each is visible in the sky for a few months before and after opposition.

The table below lists oppositions of Jupiter and Saturn between 2010 and 2031.

Table 6. Oppositions of Jupiter and Saturn, 2010–2031

| Jupiter Opposition Date | Saturn Opposition Date |
| --- | --- |
| 2010 Sep 21 | 2010 Mar 22 |
| 2011 Oct 29 | 2011 Apr 03 |
| 2012 Dec 03 | 2012 Apr 15 |
| ———— | 2013 Apr 28 |
| 2014 Jan 05 | 2014 May 10 |
| 2015 Feb 06 | 2015 May 23 |
| 2016 Mar 08 | 2016 Jun 03 |
| 2017 Apr 07 | 2017 Jun 15 |
| 2018 May 09 | 2018 Jun 27 |
| 2019 Jun 10 | 2019 Jul 09 |
| 2020 Jul 14 | 2020 Jul 20 |
| 2021 Aug 20 | 2021 Aug 02 |
| 2022 Sep 26 | 2022 Aug 14 |
| 2023 Nov 03 | 2023 Aug 27 |
| 2024 Dec 07 | 2024 Sep 08 |
| ———— | 2025 Sep 21 |
| 2026 Jan 10 | 2026 Oct 04 |
| 2027 Feb 11 | 2027 Oct 18 |
| 2028 Mar 12 | 2028 Oct 30 |
| 2029 Apr 12 | 2029 Nov 13 |
| 2030 May 13 | 2030 Nov 27 |
| 2031 Jun 15 | 2031 Dec 11 |

# Using a Planisphere

A planisphere is my favorite device for determining how the night sky will look at any date and time. The principal advantage of a planisphere over a printed star map is that it can be easily adjusted for any time on any date. Planispheres are also powerful tools for learning how the sky changes during the night and during the year.

We are all familiar with the daily motion of the Sun: It rises in the east and sets in the west. If you watch the Moon for a few hours, you will see that it, too, rises in the east and sets in the west every day. Of course, it is not the Sun and Moon that are moving. Their apparent daily motion is due to the rotation of the Earth from west to east. Since we do not usually think of ourselves as moving, we picture the sky as if it were rotating around us from east to west. If the Sun and Moon appear to be moving due to the rotation of the Earth, then the fixed stars will also appear to rotate. If you watch the sky for a few hours, you will see that the stars rise and set just like the Sun and Moon. Thus, the stars and constellations visible in the night sky are changing hour by hour.

You may also be familiar with the seasonal change of the constellations. The constellations in the sky tonight are not the same constellations that we saw in the sky 6 months ago. As the Earth revolves around the Sun, the side of the Earth facing away from the Sun is pointing in different directions at different times of the year. For example, in the Northern Hemisphere in winter, the constellation Orion is high in the sky after sunset. In summer, Orion is not visible, while the stars of Lyra, Cygnus, and Aquila are high in the sky. The summer Sun is in the same part of the sky as Orion, so Orion is up during the daytime. Thus, the constellations change through the seasons as well.

## History

The modern planisphere is derived from an ancient device called the astrolabe. The astrolabe is a 2-dimensional projection of the celestial sphere onto a plane. Its origins date back to ancient Greece, when the great astronomers Hipparchus (fl. 1$^{st}$ century B.C.) and Claudius Ptolemaeus (a.k.a. Ptolemy; c. 100–c. 170) explored and refined the projections required to plot the spherical celestial sphere on a flat plate. Theon of Alexandria (fl. 4th century A.D.) wrote a treatise on the astrolabe, and a device may have been constructed around this time. By the 6th and 7th centuries A.D., we have written descriptions of astrolabes, and the oldest extant examples are Arabic astrolabes from the 10th century A.D.

An astrolabe, a forerunner to the planisphere and one of the first devices developed to predict the appearance of the night sky.

A typical astrolabe consisted of three or more parts: a rotating star map called the rete (rhymes with treaty), one or more plates that had lines of azimuth and altitude engraved on them, and a housing called the mater. Although the rete was a thin plate, most of it was cut away leaving only a ring for the celestial equator, a ring for the ecliptic, and about 2 dozen pointers that represented the locations of bright stars. Through the fretwork rete, you could read the altitude-azimuth plate underneath. The altitude-azimuth plate was plotted for a single latitude on Earth; therefore, some astrolabes included multiple plates so that they could be used at a variety of latitudes. Most astrolabes also included a rotating arm on the back, called the alidade, for measuring the altitudes of celestial objects.

Like a planisphere, an astrolabe could show the sky for any date and time. They were primarily used as mechanical calculators for computing the time of day or night, sunrise and sunset, and the positions of the Sun and stars in the sky as well as for educational purposes. For example, to determine the time of night, an astronomer would use the alidade to measure the altitude of a star. He would rotate the rete until the star pointer for that star was located at the correct altitude, and then he could read off the time when the star had that altitude for that date. Astrolabes were used until the 17th century, when the development of spherical trigonometry and ever better instruments, such as mechanical clocks, octants, and sextants, rendered them obsolete.

Planispheres, unlike astrolabes, are not designed for precision work, so they do not include lines of altitude and azimuth drawn on the horizon plate and can be made of inexpensive materials. Modern planispheres became popular in the mid-19th century when a number of publishers produced ornate cardboard planispheres such as the Whitall Movable Planisphere (1856) and the Barrett-Serviss Star and Planet Finder (1905).

### Planisphere Layout

Most planispheres consist of two plates: a rotating base plate and an outer horizon plate. The following description refers specifically to David Chandler's The Night Sky planisphere, although most of the discussion would apply to any planisphere. A star map is printed on the base plate, with the days of the year printed along the outer edge. Brighter stars are plotted with larger symbols, while fainter stars are plotted with smaller symbols. Some of the brightest stars are labeled with their names. Printed in all capitals on the base plate are the names of the major constellations. The constellation figures are represented by lines connecting the stars. The Milky Way is seen as a shaded band cutting across the planisphere.

As you rotate the base plate, you will note that it is attached to the horizon plate by a grommet in the center. This grommet represents the North Celestial Pole. The star Polaris is located near, but not exactly at, the North Celestial Pole. On the back side of the planisphere, which shows the southern sky, the grommet represents the South Celestial Pole.

Also printed and labeled on the base plate are the celestial equator and ecliptic. The celestial equator is the Earth's equator projected out into space. It divides the celestial sphere into a northern half and a southern half. The ecliptic, which is drawn as a dashed line, marks the annual path of the Sun through the sky. As the Earth revolves (orbits) around it, the Sun appears projected against different background constellations. The equator and the ecliptic mark the origin of the celestial coordinate system. Just as we measure position on the Earth using longitude and latitude, we measure positions in the sky using right ascension (east-west) and declination (north-south). Like latitude, declination measures the north-south position of an object on the celestial sphere. The celestial equator is defined to have a declination of 0°, while the North Celestial Pole is at +90° and the South Celestial Pole is at −90°.

Right ascension, which is similar to longitude on Earth, measures the east-west position of an object in the sky. The origin for right ascension is the vernal equinox, where the Sun crosses the equator on or around March 20 of every year on its way from the Southern Hemisphere to the Northern Hemisphere. Right ascension is measured in hours, minutes, and seconds of time. On the planisphere base plate, lines of right ascension can be seen radiating from the grommet at the center. The lines of right ascension are labeled where they meet the celestial equator. Along these lines are tick marks for every 10° of declination.

The horizon plate has a clear window that allows you to see only part of the base plate at a time and has the hours of the day printed along the outer edge. The clear window represents the part of the sky that is above your horizon and is visible, and the edge of the clear window represents the horizon. The horizon plate is plotted for a specific latitude on Earth, but since planispheres are not designed for precise work, they can be used over a range of latitudes. For example, a planisphere plotted for 35° north latitude might have listed on it a range of 30–40°. The horizon is exact for a latitude of 35° and is less precise as you move farther away. Even if you are outside the listed range, within say 10-20° of the design latitude, the planisphere will still show you which stars are high in the sky for a given date and time, but the stars low around the horizon will not be accurately represented. For example, on a a planisphere plotted for 35° north latitude, if you rotate the base plate, you will see the star Canopus, the second brightest star in the sky, appear over the southern horizon for a short time. Nevertheless, Canopus is at a declination of −52.7°, so it is not visible if you live north of a latitude of 37.3°. An observer at 40° north would never see Canopus in the sky, even though the planisphere is listed as covering the range 30–40°.

Printed around the horizon window are the cardinal directions: north, south, east, and west. Due north is directly below the central grommet, while due south is exactly opposite (and the southern horizon is on the back). You may have noticed that east and west are reversed. Unlike a terrestrial map, which is meant to be viewed from above, a planisphere is meant to be held overhead. Holding a planisphere above your head will reverse east and west. Due east and due west, are located where the celestial equator meets the horizon. At all points on Earth, except the North and South Poles, the celestial equator crosses the horizon due east and due west.

The hours of the day and night are indicated along the outside edge of the horizon plate. The times from 6 pm to 6 am are labeled with a small triangle indicator. The hours of the day are also indicated on the horizon plate with small triangles, but they are not labeled.

### Activities
To see what stars are visible at a given time and date, rotate the base plate so that the date on the base plate lines up with the time on the horizon plate. For example, if you want to see what the sky looks like on January 15 at 10 pm, rotate the base plate so that January 15 lines up with the arrow for 10 pm. In the horizon window, you should see Orion at the top of the planisphere, high in the southern sky. Just rising above the eastern horizon are the stars of Leo, and just setting near the western horizon are the stars of Pegasus.

If you are on daylight saving time (or summer time), remember to subtract an hour when dialing in the time on a planisphere. That is, to see how the sky will look on July 15 at 10 pm, you should set the planisphere for July 15 at 9 pm because your watch is 1 hour ahead of standard time. To use a planisphere like an ancient astrolabe to tell time, go outside at night and set the window so that the stars in the window appear exactly as they do in the night sky. The current time will be the time that is adjacent to the current date.

To use your planisphere to determine when an object will rise, you can rotate the base plate and move the object until it is on the eastern horizon (for objects south of the equator, you should use the back of the planisphere). The time that the object rises on any date is the time located next to that date. For example, to determine when Betelgeuse will rise on October 15, rotate the base plate until Betelgeuse is on the eastern horizon. Find October 15 on the edge of the base plate, and you will see that Betelgeuse rises at 9:54 pm. You can determine when Betelgeuse will set by moving it over to the western horizon and reading the time next to the date.

You can locate the Sun on your planisphere by recalling that the Sun is always along the ecliptic in the sky. On or around the vernal equinox, about March 20 of each year, the Sun crosses the equator as it heads from the Southern Hemisphere to the Northern Hemisphere. This point is located in the constellation Pisces and defines the origin point ($0^h$) of right ascension. Thus, around March 20, the Sun has a right ascension of $0^h$. The Sun moves about 2 hours of right ascension each month, or about 30 minutes each week. Also keep in mind the other 3 marker points of the year. On or around June 21, about a quarter of the year has passed since the vernal equinox, so the Sun will have completed a quarter of its path around the sky. Thus, the Sun will be at $6^h$ right ascension on the ecliptic, very near the object M35 in the constellation Gemini. On the autumnal equinox, around September 22, the Sun will be at $12^h$ right ascension, crossing the equator in the constellation Virgo. On the winter solstice, around December 21, the Sun will be at $18^h$ right ascension in Sagittarius near the object M8 on your planisphere.

Now we can use the position of the Sun to determine sunrise and sunset times by placing the Sun on the horizon. For example, let's calculate, roughly, when the Sun will rise on August 21. That date is about 2 months after the summer solstice (and 1 month before the autumnal equinox), so the Sun will be at a right ascension of about $10^h$, and it will be along the ecliptic. After finding the ecliptic on the planisphere, trace it around until you come to the part of the ecliptic that would have a right ascension of $10^h$. This is close to the star Regulus in Leo. Now rotate that point on the ecliptic to the eastern horizon. You will see that the Sun will rise at about 5:20 am. Since in most of North America, we are on daylight saving time in August, I need to add an hour to see that sunrise will be around 6:20 am that day. Your estimated time of sunrise is only accurate for the center of your time zone. If you are east of the center of your time zone, the Sun could rise as much as 30 minutes earlier. If you are west of the center of your time zone, it could rise as much as 30 minutes later. ■

# The Best Annual Meteor Showers

This table lists the 10 best meteor showers of the year. The meteors in a meteor shower appear to radiate out of a constellation, and the constellation in which this **radiant** point is located gives the shower its name. The dates listed in the table are the dates over which the shower is active. In most cases, however, the shower is best on a single night. The date of the maximum can vary by a day or two from year to year. Be sure to look online or in an astronomy magazine for the dates of maxima in a given year. Some showers put on a good show for only a few hours or less. The Quadrantids in January, for example, usually have a peak that lasts about 1 hour. All the showers listed are visible from the mid–Northern Hemisphere. Those whose declanation is less than +50° are visible from the mid–Southern Hemisphere as well.

The **zenith hourly rate** is the number of meteors per hour that you would expect to see under very dark, completely transparent skies if the radiant of the shower is high overhead at the zenith and the Moon is not in the sky. Since many of us watch meteor showers from moderately light-polluted skies that are less than fully transparent and the radiant is rarely high overhead, most showers produce less than the listed maximum. The table also lists the right ascension and declination of the radiant and the constellation in which it is located. To best see a meteor shower, the radiant point must be above the horizon. For most meteor showers, this means they are best observed after midnight and in the early-morning hours before sunrise. The parent body that generates the small meteoroids is listed in the last column. Most showers originate from comets, but some, such as the Geminids, come from asteroids. ■

## Important Terms

**radiant**: A point from which a meteor shower seems to originate.

**zenith hourly rate**: A measure of how many meteors in a meteor shower can be seen per hour at the zenith.

**Table 7. The best annual meteor showers**

| Shower | Dates | Date of Maximum | ZHR | Radiant | Constellation | Parent Body |
|---|---|---|---|---|---|---|
| Quadrantids | 1–6 Jan | 3 Jan | 60 | $15^h30^m$ +50 | Boötes | Minor planet 2003 EH1 |
| Lyrids | 19–25 Apr | 22 Apr | 10 | $18^h10^m$ +32 | Lyra | Comet Thatcher |
| η Aquarids | 1–10 May | 6 May | 35 | $22^h23^m$ −01 | Aquarius | Comet Halley |
| δ Aquarids | 15 Jul–15 Aug | 29 Jul–7 Aug | 20 10 | $22^h39^m$ −17 $23^h07^m$ +02 | Aquarius Pisces | unknown |
| Perseids | 23 Jul–20 Aug | 12 Aug | 80 | $03^h08^m$ +58 | Perseus | Comet Swift-Tuttle |
| Orionids | 16–27 Oct | 20–22 Oct | 25 | $06^h27^m$ +15 | Orion/Gemini | Comet Halley |
| Taurids | 20 Oct–30 Nov | 3 Nov | 10 | $03^h47^m$ +14 and +22 | Taurus | Comet Encke |
| Leonids | 15–20 Nov | 17 Nov | 15 | $10^h11^m$ +22 | Leo | Comet Temple-Tuttle |
| Geminids | 7–16 Dec | 13 Dec | 100 | $07^h28^m$ +32 | Gemini | Minor planet Phaethon |
| Ursids | 17–25 Dec | 22 Dec | 10 | $14^h27^m$ +78 | Ursa Minor | Comet Tuttle |

*Note*: ZHR = zenith hourly rate. Radiant is given as right ascension, followed by declanation.

# Glossary

**accretion disk**: A rotating disk of gas and dust that forms around the opening of a black hole.

**altitude**: The angular distance between a celestial object and the closest point on the horizon. The horizon has an altitude of 0°, while the zenith has an altitude of 90°.

**altitude-azimuth mount**: A telescope mount that moves in altitude (up-down) and in azimuth (left-right).

**aperture**: The diameter of a telescope's open end that allows light to enter.

**apparent brightness**: The amount of light (or energy) received on Earth from a distant celestial source. The brightness depends on both the luminosity of the source and the distance to the source.

**apparent magnitude**: A system for measuring the apparent brightness of an object, developed by Hipparchus around 150 B.C.

**arc minute**: A unit used to measure angles; there are 60 minutes of arc (60′) in 1°.

**arc second**: A unit used to measure angles; there are 60 seconds of arc (60″) in an arc minute (1′) and 3600 arc seconds in 1°.

**asterism**: An easily recognizable pattern of stars useful for navigating the sky. Asterisms are not officially recognized.

**astronomical unit (AU)**: The average distance from the Earth to the Sun, about 150 million km or 93 million miles.

**averted vision**: A technique for making faint objects more noticeable by looking near, but not directly at, the object.

**azimuth**: The angle between the position of a celestial object and true north, measured through east. That is, north is at an azimuth of 0°, east is 90°, south is 180°, and west is 270°.

**Barlow lens**: A lens that increases the focal length of a telescope, thus increasing the magnification of any eyepieces used with it.

**black hole**: A compact, massive object whose gravity is so strong that not even light can escape from nearby.

**cassegrain**: A telescope in which the secondary mirror diverts the light through a hole in the primary mirror and out the bottom of the telescope.

**catadioptric**: A telescope that uses both lenses and mirrors to bring light to a focus.

**celestial equator**: The Earth's equator projected onto the celestial sphere. The celestial equator divides the sky into a northern half and a southern half.

**celestial poles**: The points on the celestial sphere directly above the Earth's North and South poles. The sky appears to rotate around the North and South celestial poles, moving from east to west every day as the Earth rotates.

**celestial sphere**: An imaginary sphere surrounding the Earth that contains the Sun, Moon, planets, and stars. Although the celestial sphere is not real, it is useful for thinking about the motions of the objects in our sky.

**Cepheid variable stars**: A class of giant, unstable stars that vary in brightness. Their luminosity is related to the length of time it takes the star to vary in brightness, making them useful stars for measuring distances.

**chromatic aberration**: An aberration found in refracting telescopes where not all colors of light can be brought to a focus at the same time.

**circumpolar**: Any star or constellation that circles around one of the celestial poles, never rising or setting. Whether a given constellation is circumpolar depends on your latitude. At the North and South poles, all constellations are circumpolar. At the equator, no constellations are circumpolar.

**conjunction**: A pair of celestial bodies having the same right ascension. During a conjunction, the two objects appear close together in the sky.

**constellation**: One of 88 bounded regions in the sky adopted by the International Astronomical Union in 1922. Before then, a constellation was a group of stars that pictured a mythological person, animal, or object.

**corona**: The hot outer atmosphere of the Sun only visible on Earth during a total solar eclipse.

**declination (DEC)**: The north-south position of an object on the celestial sphere. Declination is measured with respect to the celestial equator. An object north of the equator has a positive declination, while an object south of the equator has a negative declination. Declination is measured in degrees, arc minutes, and arc seconds.

**Dobsonian telescope**: A reflecting telescope on an altitude-azimuth mount made of inexpensive materials.

**double or multiple stars**: Two or more stars in orbit around one another.

**ecliptic**: The annual path of the Sun through the sky.

**equatorial mount**: A telescope mount in which one axis of movement is aligned with the rotation axis of the Earth (i.e., one axis points at the celestial pole). To track the stars, the telescope is moved in one axis only at a constant rate.

**equinoxes**: The days when the Sun crosses the celestial equator every year, occurring around March 20 and September 22.

**eye relief**: The maximum distance that an observer can place his or her eye from an eyepiece and still see the full field. Small eye relief means that an observer's eye must be very close to the eyepiece, making it difficult for people with eyeglasses to find a comfortable viewing position.

**galaxy**: A large group of hundreds of millions to as many as a trillion stars held together by their mutual gravity.

**globular cluster**: A cluster of stars that contains between 10,000 and a few million stars. Such clusters are primarily found in the halo of our Milky Way galaxy. Most globular clusters are old, and therefore they contain mostly old yellow, orange, and red stars.

**greatest elongation**: The point in the orbit of a planet closer to the Sun than the Earth when it has its maximum angular distance from the Sun.

**heliacal rising**: The rising of a celestial body, such as a star, before the Sun.

**horizon**: The boundary where the sky meets the earth. The horizon blocks half of the celestial sphere from view.

**Jovian planets**: Giant gaseous planets in the middle solar system: Jupiter, Saturn, Uranus, and Neptune.

**Kuiper belt**: A region of cold, icy bodies at the edge of our solar system.

**light pollution**: The illumination of the night sky by stray light from human activities.

**light-year**: The distance that light travels in 1 year, about 9.5 trillion kilometers or 5.9 trillion miles.

**limiting magnitude**: An estimate of the darkness of your night sky found by noting the magnitude of the faintest stars that you can see at the zenith.

**luminosity**: The total energy released every second by a star or other celestial object.

**magnitude system**: A way that astronomers describe the apparent brightness of a star. In this system, a difference of five magnitudes is exactly a factor of 100 in brightness.

**meridian**: A line running from due north to due south that passes through your zenith. Your meridian cuts the sky into an eastern half and a western half.

**meteor**: A bright streak of light in the sky that appears when a meteoroid enters the Earth's atmosphere.

**nebula**: A cloud of interstellar dust or gas.

**Newtonian**: A reflecting telescope in which a flat secondary mirror diverts the light out the side of the tube near the top of the telescope.

**North Celestial Pole**: The point directly above the Earth's North Pole projected into space.

**objective**: A term used to describe the primary focusing optical element (lens or mirror) in a telescope.

**Oort cloud**: A giant cloud of cold, icy bodies that stretches tremendous distances away from the Sun.

**open cluster**: A cluster of stars that contains between a few dozen and a few thousand stars, including many young, hot, blue stars. Open clusters are found primarily in the plane of the Milky Way.

**opposition**: The time when planets farther from the Sun than the Earth are exactly opposite the Sun in the sky. Opposition occurs when the Earth passes between the planet and the Sun. At this time, the planet is closest to Earth, appears largest in a telescope, and is high in the sky at midnight.

**parallax**: A technique for measuring the distances to stars by watching them shift back and forth in the sky due to the Earth's revolution around the Sun.

**penumbra**: The part of the shadow of a body where the Sun is only partially blocked.

**planetary nebula**: A cloud of gas shed by a low-mass star as it is dying.

**planisphere**: A star map that consists of a rotating disk that can be set to show the night sky for any day or time.

**precession**: The 26,000-year wobble of the Earth's rotation axis due to the gravitational attraction of the Moon. Due to precession, the North and South celestial poles are shifting in the sky and the vernal equinox point is regressing through the zodiacal constellations.

**prominences**: Loops of gas extending up from the surface of the Sun, best seen when they are on the edge of Sun.

**quasar**: The bright, starlike nucleus of a galaxy containing a supermassive black hole in which one of the jets of material from the black hole is aimed in our direction.

**radiant**: A point from which a meteor shower seems to originate.

**red giant**: A low- to moderate-mass star in the last stages of its lifetime; such a star has fused all the hydrogen atoms at its core into helium atoms.

**reflector**: A telescope that uses a mirror to bring light to a focus.

**refractor**: A telescope that uses a lens to bring light to a focus.

**right ascension (RA)**: The east-west position of a celestial object in the sky. Right ascension is measured from the vernal equinox toward the east. It is measured in hours, minutes, and seconds of time.

**seeing**: An estimate of the steadiness of the night sky during an observing session, which can be estimated by watching the twinkling of the stars. If stars high overhead are twinkling, then the atmosphere is unsteady, the seeing is bad, and you will not see many details through a telescope. If the stars high overhead *and* those near the horizon are not twinkling, then the atmosphere is steady, seeing is good, and you will see finer details in a telescope.

**solstices**: The times when the Sun reaches its maximum northern or southern declination every year (23.5°), occurring around June 21 and December 21.

**star atlas**: A detailed map or maps of the sky showing stars and other objects in their accurate positions, meant to be used when trying to find objects with binoculars or a telescope. A star atlas typically plots the whole sky. It does not indicate which stars or objects are above the horizon. Due to precession, star atlases are plotted for a standard year, called the epoch, which is used for the following 20–30 years.

**star diagonal**: A right-angle mirror or prism that diverts the beam of light from a telescope to the side, making it easier for an observer to position his or her head to look through the telescope.

**star map**: A less-detailed map of the sky than a star atlas that shows only the brightest stars and other bright objects that are above the horizon. Star maps are plotted for a specific latitude and apply for only certain dates and times.

**sundogs** (a.k.a. **parhelia**): Bright spots of light 22° to the left and/or right of the Sun created by small ice crystals in high, thin cirrus clouds.

**sunspot**: A magnetic storm on the surface of the Sun that appears darker than the rest of the surface.

**supernova**: The explosion of a massive star at the end of its life.

**supergiant**: A massive star of great luminosity.

**telescope**: An instrument for viewing distant objects; telescopes come in three basic types: refractor, reflector, and catadioptric.

**terminator**: The boundary between day and night on a celestial body.

**terrestrial planets**: Rocky planets in the inner solar system: Mercury, Venus, Earth, and Mars.

**totality**: The period during a solar eclipse when the Moon completely covers the Sun. **transit**: When one planet crosses directly in front of the Sun as seen from another planet.

**Tropic of Cancer**: The latitude on Earth around 23.5° north where the Sun is directly overhead at noon on the summer solstice (around June 21).

**Tropic of Capricorn**: The latitude on Earth around 23.5° south where the Sun is directly overhead at noon on the winter solstice (around December 21).

**twilight**: The time between sunset and full darkness. Astronomers define three kinds of twilight: civil, from sunset to when the altitude of the Sun is 6° below the horizon (−6°); nautical, when the Sun is 6° to 12° below the horizon (−6° to −12°); and astronomical, when the Sun is 12° to 18° below the horizon (−12° to −18°). At the end of astronomical twilight, the sky is dark enough to observe faint objects.

**umbra**: The dark, cone-shaped part of the shadow of a body transiting the Sun; to an observer in the umbra, the Sun is completely blocked from view.

**visual double star**: A binary star in which each of the two components is visible in a telescope.

**waxing**: The phase of the Moon during the transition from new to full.

**white dwarf star**: The burned-out core of a low- to moderate-mass star.

**zenith**: The point on the celestial sphere straight overhead.

**zenith hourly rate**: A measure of how many meteors in a meteor shower can be seen per hour at the zenith.

**zodiac**: The band of 12 classical constellations through which the Sun passes in its annual motion around the sky on the ecliptic. Since the planets and the Moon have orbits nearly aligned with the ecliptic, they too are usually found in the zodiacal constellations. With the definition of the modern constellation boundaries, Ophiuchus is now the thirteenth zodiacal constellation.

# Bibliography

The books in this bibliography fall into 6 categories: star atlases, books on the mythology and history of the constellations, guides to viewing the sky, books on astronomy, books on observing special events, and books on equipment and telescopes.

A good star atlas is an essential purchase if you plan to observe the sky with binoculars or a telescope. I strongly recommend *Norton's Star Atlas,* edited by Ian Ridpath. *Norton's* is on the shelf of nearly every backyard astronomer. In addition to including great star maps, the book is an excellent reference guide on the night sky, buying and using equipment, and observing the Sun, Moon, and planets. If you need an atlas with more detail and one that plots fainter stars, I recommend *Sky Atlas 2000.0* by Wil Tirion and Roger Sinnott. Its large format and spiral binding make it easy to use at night. For finding the faintest objects under dark skies with a large telescope, I recommend the 2-volume *Uranometria 2000.0* by Wil Tirion, Barry Rappaport, and Will Remaklus.

If you are interested in the mythology of the constellations, I recommend that you start with *The New Patterns in the Sky: Myths and Legends of the Stars* by Julius Staal. It includes not only the Greek and Roman legends but stories from other cultures as well. For a better accounting of the classical Greek stories and a short history of the constellations, I recommend *Star Tales* by Ian Ridpath. For a fascinating accounting of the origins of the proper names of our stars, see *A Dictionary of Modern Star Names* by Paul Kunitzsch and Tim Smart.

An excellent guide to the night sky is *A Year of the Stars: A Month-by-Month Journey of Skywatching* by Fred Schaaf. You should also consider subscribing to one or more of the monthly magazines for amateur astronomers. In North America, the most popular are *Sky & Telescope* and *Astronomy*. Every month, the magazines provide a map of the sky, articles on the objects that are visible, equipment reviews, and articles on the latest developments in astronomy. I strongly recommend that you read the magazines for a few months before you purchase a telescope. The articles and advertisements will help you learn the jargon and become familiar with the types of equipment that are in use. For first-time telescope users, I highly recommend *Turn Left at Orion* by Guy Consolmagno and Dan Davis.

Finally, I also recommend the classic 3-volume *Burnham's Celestial Handbook: An Observer's Guide to the Universe Beyond the Solar System* by Robert Burnham Jr. These books contain a wealth of information on all the interesting objects visible through a backyard telescope, arranged by constellation. Although some of the factual data (such as distances) are out of date, the majority of the content is still relevant. It is a great resource for learning more about the nature and significance of the objects you are observing.

---

Allen, Richard Hinckley. *Star Names: Their Lore and Meaning.* Mineola, NY: Dover Publications, 1963. This reprint of the classic 1899 book *Star-Names and Their Meanings* by Allen provides a wealth of information on the origins of the names of the stars and stories of the constellations from the myths and legends of many cultures. Nevertheless, some of Allen's translations and interpretations have been called into question by modern scholars (see Kunitzsch and Smart for a more widely accepted reference ).

*Astronomy* magazine. Waukesha, WI: Kalmbach Publishing.

Brunier, Serge, and Jean-Pierre Luminet. *Glourious Eclipses: Their Past, Present, and Future.* New York: Cambridge University Press, 2000. An inspiring account of historical and modern efforts to observe total solar eclipses.

Burnham, Robert Jr., *Burnham's Celestial Handbook: An Observer's Guide to the Universe Beyond the Solar System.* Mineola, NY: Dover Publications, 1978. Like Norton's *Star Atlas*, this 3-volume classic is a must-have for any astronomer. It is a magnificent compilation of information on double stars, deep-sky and other interesting objects, mythology, astronomy, and even some philosophy, all arranged by constellation. Before a night of observing, be sure to read a few entries for the constellations that are above the horizon that night. It will greatly increase your appreciation of what you are seeing and explain why it is significant. Though some of the factual data in this book is out of date, the vast majority is timeless.

Chandler, David S. *The Night Sky Planisphere.* Springville, CA: The David Chandler Company, 1992. Chandler's planisphere is one of the best available. Versions exist for 4 latitude ranges in the Northern Hemisphere and 1 in the Southern Hemisphere. To reduce distortion in their shapes, the Southern Hemisphere's constellations are found on the back of the Northern Hemisphere versions.

Condos, Theony. *Star Myths.* Grand Rapids, MI: Phanes Press, 1997. Condos provides English translations of, and commentaries on, 2 of the most important ancient sources of our classical Western myths associated with the constellations: *Catasterismi (Constellations)* by Eratosthenes and *Poeticon Astronomicon (The Poetic Astronomy)* by Hyginus.

Consolmagno, Guy, and Dan Davis. *Turn Left at Orion: A Hundred Night Sky Objects to See in a Small Telescope—and How to Find Them.* 3rd ed. New York: Cambridge University Press, 2000. An excellent introduction to finding objects with a backyard telescope. This book is especially good for a first-time telescope owner.

Covington, Michael A. *Astrophotography for the Amateur.* 2nd ed. New York: Cambridge University Press, 1999. If you are interesting in photographing the sky, either with or without a telescope, this book is a great guide to getting started. It includes sections on all aspects of astrophotography, such as taking star-trail images without a telescope; photographing the Moon, planets, and eclipses; and taking hours-long exposures on deep-sky objects. It also includes very good chapters on equipment, film, and digital astrophotography as well as image processing for those using digital cameras.

Cragan, Murray, and Emil Bonnano. *Uranometria 2000.0: Deep Sky Atlas.* Richmond, VA: Willmann-Bell, 2001. This companion to the 2-volume *Uranometria* star atlases by Tirion, Rappaport, and Remaklus provides coordinates, diameters, orientations, magnitudes, types, and notes for each of the deep-sky objects plotted on the star maps. A very useful reference for learning more about the objects you are observing.

Dickinson, Terrance. *Nightwatch: A Practical Guide to Viewing the Unvierse*. 4<sup>th</sup> ed. Buffalo, NY: Firefly Books, 2006. A very nice introduction to observing the night sky for beginners in the Northern Hemisphere. It includes all sky maps for finding the constellations and somewhat more detailed maps for binocular and small telescope observing.

Edmund Scientific. *Edmund Mag 5 Star Atlas*. Barrington, NJ: Edmund Scientific, 1989. An inexpensive set of all-sky star maps with all stars down to magnitude 5.0. Though these star maps do not plot nearly as many stars as some of the other atlases listed here, it is a great atlas for beginners due to its simplicity.

Evans, James. *The History and Practice of Ancient Astronomy*. New York: Oxford University Press, 1998. This book is for those interested in learning more about how the ancients practiced astronomy. From designing sundials to calculating the position of the Sun, Moon, and planets, this work will show you how it was done from ancient times until the Renaissance.

Heifetz, Milton D., and Wil Tirion. *A Walk through the Heavens*. 3<sup>rd</sup> ed. New York: Cambridge University Press, 2004. This book contains a series of star maps and step-by-step directions for finding and observing the northern constellations with the naked eye. It includes a short summary of the mythology of each of the constellations.

———. *A Walk through the Southern Skies*, 2<sup>nd</sup> ed. New York: Cambridge University Press, 2007. A companion to *A Walk through the Heavens,* this book has star maps and step-by-step directions for finding and observing the southern constellations with the naked eye. It also includes some mythology for each of the constellations.

Hirshfeld, Alan W. *Parallax: The Race to Measure the Cosmos*. New York: Henry Holt, 2001. This book provides an engaging account of astronomers' efforts to measure the distances to the nearby stars by using parallax.

Kaler, James B. *The Ever-Changing Sky*. New York: Cambridge University Press, 1996. A very good introduction to the celestial sphere and the motions of the Sun, Moon, and planets on it.

————. *Stars and Their Spectra*. New York: Cambridge University Press, 1997. If you would like to know more about the stars in our night sky and how astronomers determine their basic properties, such as temperature, luminosity, mass, and so forth, this book is an excellent starting point.

Kunitzsch, Paul, and Tim Smart. *A Dictionary of Modern Star Names*. Cambridge, MA: Sky Publishing, 2006. This short guide provides an authoritative account of the derivation of the names of 254 of the brightest stars in our night sky, arranged alphabetically by constellation.

Levy, David H. *David Levy's Guide to the Night Sky*. New York: Cambridge University Press, 2001. A very thorough introduction to observing the night sky, including a good discussion of equipment. My favorite part of the book is a tour of the features on the Moon visible night by night as the Moon goes through its monthly phases.

Littmann, Mark, Ken Willcox, and Fred Espenak. *Totality: Eclipses of the Sun*. 2nd ed. New York: Oxford University Press, 1999. If you are interested in traveling to see a total solar eclipse, you must read this book. It provides not only a wonderful summary of the history, mythology, and mechanics of total solar eclipses but a thorough observing guide to help you plan your few minutes inside the shadow of the Moon.

Maor, Eli. *Venus in Transit*. Princeton, NJ: Princeton University Press, 2004. A very good introduction to understanding the rare times that Venus transits the Sun.

Maryboy, Nancy C., and David Begay. *Sharing the Skies: Navajo Astronomy*. Bluff, UT: Indigenous Education Institute, 2005. This book contains a summary of traditional Navajo astronomy and cosmology.

Millar, William. *The Amateur Astronomer's Introduction to the Celestial Sphere*. New York: Cambridge University Press, 2006. This book is an outstanding introduction to the layout and motions of the celestial sphere. It includes discussions on the motions of the Sun, Moon, stars, and planets in our night sky and excellent sections on celestial coordinate systems, timekeeping, phases of the Moon, the seasons, eclipses, and more.

Monroe, Jean Guard, and Ray A. Williamson. *They Dance in the Sky: Native American Star Myths*. Boston: Houghton Mifflin, 1987. A very nice introduction to the sky lore of various Native American tribes and what these stories tell us about their cultures.

*The Observer's Handbook*. An annual publication of the Royal Astronomical Society of Canada, *The Observer's Handbook* lists and details all of the astronomical events for the coming year, plus it contains a huge amount of other useful information.

Olcott, William Tyler. *Star Lore: Myths, Legends, and Facts*. Mineola, NY: Dover Publications, 2004. This is a reprint edition of a 1911 classic compilation of constellation myths and legends from one of the greatest astronomy writers of the early 20th century. After you have read Ridpath's *Star Tales* or Staal's *The New Patterns in the Sky* for a good introduction to the star lore, Olcott's book will provide a more thorough accounting of the history of the stories and the constellations.

O'Meara, Stephan James. *Deep-Sky Companions: The Caldwell Objects*. Cambridge, MA: Sky Publishing, 2002. After you have exhausted the Messier objects (see below), you can start on the Caldwell list. This excellent guide contains information on finding and seeing deep-sky objects that did not make it onto Messier's list.

————. *Deep-Sky Companions: The Messier Objects*. Cambridge, MA: Sky Publishing, 1998. This is an excellent guide to observing the brightest of the deep-sky objects. Not only does it include useful information on finding and observing the objects; it provides a nice summary of the physical nature of each object and why it is interesting.

Pasachoff, Jay M. *A Field Guide to the Stars and Planets*, Peterson Field Guides. New York: Houghton Mifflin, 2000. An all-in-one introduction to astronomy with both all-sky maps for learning the constellations and deep-sky maps for using a telescope. Unfortunately, the small format of the book, and the fact that its binding will not open flat, makes the charts difficult to use at night.

Ridpath, Ian. *Star Tales*. New York: Universe Books 1988. Ridpath has written an engaging account of the development and history of the modern constellations with a thorough compilation of the of the classical Western (Greek and Roman) myths associated with them.

————, ed. *Norton's Star Atlas*. 20th ed. New York: Pi Press, 2004. If you can buy only one astronomy book, this is it. In addition to a great set of all-sky charts, the book contains a wealth of information on using telescopes and observing the sky. Its enduring popularity is evidenced by the fact that the book is in its 20th edition; nevertheless, the publisher has kept the book up to date with recent advancements in telescopes and accessories.

Rukl, Antonin. *Sky & Telescope's Field Map of the Moon*. Cambridge, MA: Sky Publishing, 2007. A very good map of the Moon that also comes in a mirror-image version for use at the eyepiece if your telescope produces a mirror-reversed image.

Rukl, Antonin, and Gary Seronik. *Atlas of the Moon*. Rev. ed. Cambridge, MA: Sky Publishing, 2007. The best detailed atlas of the near side of the Moon.

Schaaf, Fred. *A Year of the Stars: A Month-by-Month Journey of Skywatching*. Amherst, NY: Prometheus Books, 2003. Schaaf has produced one of the finest month-by-month observing guides to the night sky of the Northern Hemisphere. For each month, he lists the constellations that are visible, discusses the important objects that can be seen in binoculars or a telescope, and tells us of other interesting astronomical phenomena that occur.

Sheehan, William, and John Westfall. *The Transits of Venus*. New York: Prometheus Books, 2004. An account of the history and importance of the rare transits of Venus, including a guide to observing the transit in 2012.

Sinnott, Roger W. *Sky & Telescope's Pocket Sky Atlas*. Cambridge, MA: Sky Publishing, 2006. This is a very good, compact sky atlas that plots all stars down to magnitude 7.6 and about 1500 deep-sky objects.

*Sky & Telescope* magazine. Every month, *Sky & Telescope* publishes articles on current events in astronomy, updates on astronomical research, guides to the planets and deep-sky objects visible that month, equipment reviews, and advice.

Staal, Julius D. W. *The New Patterns in the Sky: Myths and Legends of the Stars*. Blacksburg, VA: McDonald and Woodward, 1988. An excellent overview of the myths of the constellations, including not only classical Greek and Roman myths but many stories and legends from other cultures.

———. Stars *of Jade: Calendar Lore, Mythology, Legends, and Star Stories of Ancient China*. Atlanta, GA: Writ Press, 1984. A thorough accounting of ancient Chinese astronomy, constellations, and asterisms.

Stott, Carole. *Celestial Charts: Antique Maps of the Heavens*. London: Studio Editions, 1995. A beautiful picture-book survey of efforts to map and chart the night sky.

Tirion, Wil, Barry Rappaport, and Will Remaklus. *Uranometria 2000.0*. Richmond, VA: Willmann-Bell, 2001. The definitive star atlas of the Northern Hemisphere (vol. 1) and Southern Hemisphere (vol. 2) for deep-sky observers, containing all stars to magnitude 9.75 and over 30,000 deep-sky objects, and detailed charts on crowded regions of interest.

Tirion, Wil, and Roger W. Sinnott. *Sky Atlas 2000.0*. 2nd deluxe ed. Cambridge, MA: Sky Publishing, 1998. As you progress in your ability to find faint, deep-sky objects, you will need a better star atlas. This large-format atlas is one of the best, with all stars down to magnitude 8.5 and 2700 deep-sky objects.

Toomer, G. J. *Ptolemy's Almagest*. Princeton, NJ: Princeton University Press, 1998. Toomer provides here the definitive English translation of Ptolemy's

*Almagest* for those interested in seeing the original source of the classical 48 constellations and one of the most influential astronomical works in history.

Wood, Charles. *The Modern Moon: A Personal View*. Cambridge, MA: Sky Publishing, 2007. For those who are interested in observing the Moon, this is an essential guide to understanding the geology and history of the face of our nearest neighbor.

**Websites**

*Atmospheric Optics*. http://www.atoptics.co.uk. If you see a rainbow, a sundog, or a halo around the Moon, the Atmospheric Optics page will explain what is going on.

*Heavens-Above*. http://www.heavens-above.com. *Heavens-Above* is the best website for predicting when and where you can see satellites in the night sky. Before you go out for a night of observing, be sure to run the predictions to see which satellites will be visible that night and when and where to look for them. The website also includes dozens of other useful tools for predicting sunrise, sunset, moonrise, moonset, twilight times, and so forth. Make sure that you correctly set your longitude and latitude when using the program!

*Lunar Photo of the Day*. http://lpod.wikispaces.com. The *Lunar Photo of the Day* website is a great resource for those interested in observing the Moon and learning about the geology of the Moon. Like *Astronomy Picture of the Day*, a new image with explanation is loaded every day.

*NASA*. http://www.nasa.gov. This is the authoritative website for keeping up with U.S. space exploration. The space station pages provide a number of tools for predicting when the International Space Station and other NASA spacecraft will be visible in the sky.

*NASA Astronomy Picture of the Day*. http://antwrp.gsfc.nasa.gov. The *NASA Astronomy Picture of the Day* website is one of the best astronomy resources on the Internet. Every day, the authors provide an astronomical image with an explanation, including many links for finding more information.

*NASA Eclipse Page.* http://eclipse.gsfc.nasa.gov. The definitive source for information on solar and lunar eclipses and planetary transits. For each upcoming eclipse, the site provides a special bulletin that includes detailed predictions of times and locations and meteorological data to help you plan when and where to go to see an eclipse.

*Skymaps.com.* http://www.skymaps.com. If you want a very nice map of the night sky, you can print one from here. Every month, they produce maps of the night sky for both the Northern and Southern hemispheres. The maps include a monthly summary of events to look for.

*Spaceweather.com.* http://www.spaceweather.com. A great resource for watching for changing events in the sky. They provide aurora alerts and predictions, meteor shower predictions, and many amazing photos of interesting events in the sky.

*Star of the Week.* http://stars.astro.illinois.edu/sow/sowlist.html. If you want to know more about the bright stars in our sky, visit this, Jim Kaler's web page on stars. In this huge alphabetical listing of stars, you can learn about each of the bright stars in our sky, including information on its distance, its name, and its physical nature.

# Notes